EXPLORING
NATURE'S BOUNTY

One Hundred Outings near New York City

Vineyards,
Orchards,
Farms,
Nature Preserves,
Historic Herb Gardens,
U-Pick-It Sites,
Apiaries,
Greenhouses, and
Fruits and Vegetables Galore

LUCY D. ROSENFELD
MARINA HARRISON

RIVERGATE BOOKS
An imprint of Rutgers University Press
New Brunswick, New Jersey, and London

Library of Congress Cataloging-in-Publication Data
Rosenfeld, Lucy D., 1939–
Exploring nature's bounty : one hundred outings near New York City :
vineyards, orchards, farms, nature preserves, historic herb gardens, u-pick-it sites,
apiaries, greenhouses, and fruits and vegetables galore / Lucy D. Rosenfeld and
Marina Harrison.
p. cm.
Includes index.
ISBN 978-0-8135-5249-1 (pbk. : alk. paper)
1. Horticulture—New York Region—Guidebooks. 2. Farms—New York Region—
Guidebooks. 3. Gardens—New York Region—Guidebooks. 4. Horticulture—
Pennsylvania—Guidebooks. 5. Farms—Pennsylvania—Guidebooks.
6. Gardens—Pennsylvania—Guidebooks. 7. New York Region—Guidebooks.
8. Pennsylvania—Guidebooks. I. Harrison, Marina, 1939– II. Title.
SB85.N68R67 2012
635'.048097471—dc23   2011023343

A British Cataloging-in-Publication record for this book
is available from the British Library.

Frontispiece: Stone Barns cabbages in field, Pocantico Hills, NY (Courtesy
Stone Barns, Anabel Braithwaite for Belathée Photography)

Visit our Web site: http://rutgerspress.rutgers.edu

Manufactured in the United States of America

CONTENTS

NEW YORK CITY

NEW YORK STATE: LONG ISLAND

NEW YORK STATE: EAST OF THE HUDSON

NEW YORK STATE: WEST OF THE HUDSON

NEW JERSEY

PENNSYLVANIA AND DELAWARE

PREFACE

We called this book *Exploring Nature's Bounty* because in our many outings in the region we were searching for the gifts that nature offers. And, despite our fairly urban tristate area, we have been happily surprised to find such a rich array of farms, orchards, vineyards, nature preserves, and gardens filled with flourishing crops—nature's bounty.

The places we have chosen all welcome visitors; and you will be able to see firsthand the art of farming, and sometimes to pick your own produce or to help out on the farm—or simply enjoy the ambiance of the great outdoors so close to the city. These are places where you and your family can walk among the rows of burgeoning grapevines, corn stalks, apple trees, and so much more.

Our farms range from traditional fruit orchards to greenhouses filled with hydroponic (water-grown) tomatoes and basil, to neatly ordered herb gardens in historic settings. Recent interest in vineyards has spread throughout the region (especially on Long Island's North Fork), and these make for wonderful and often glamorous places to visit. Some of our venues focus on the preservation of crops, such as the American chestnut, which was nearly obliterated, while in others we introduce you to honey making and maple sugaring. Some sites offer teaching and demonstrations on agricultural methods, and many are geared to children. (We include corn mazes, hayrides, and pumpkin picking, for example.)

We recommend that you contact each site before visiting, to make sure of your timing for an outing, since nature is not always predictable; obviously it is best to go in growing or harvesting season. We include festivals featuring local produce and, at the end of the book, a guide to choosing an outing that best suits you, your family, and your taste buds. Directions are included in each write-up, beginning with the nearest major highway. We have tried to limit our venues to those within a two-hour radius of New York City (more or less).

We wish you as delightful and tasty experiences as we have had in researching this book. Feel free to send us descriptions of your adventures on these farms and any others you may have discovered along the way.

Dr. Davies Farm, Congers, NY

EXPLORING NATURE'S BOUNTY

CONNECTICUT

1 •
THE WORLD OF HERBS

The Thriving Center at Gilbertie's Herb Gardens, Westport

✧ DIRECTIONS

7 Sylvan Lane, Westport, CT. Take the Merritt Parkway to exit 41 and go left onto Route 33. In about three miles (just before the Sunoco gas station), turn right onto Sylvan Road. Follow the signs to Gilbertie's Herb Gardens.

✧ INFORMATION

Open Tuesday–Saturday, 8:30 A.M.–5:30 P.M.; Sunday, 9 A.M.–4 P.M. Telephone: 203-227-4175. Web site: www.gilbertiesherbs.com.

✧ It is a joy to discover Gilbertie's Herb Gardens in Westport. Not only do you see a sampling of the great varieties of herbs here, as well as a group of fenced-in outdoor display gardens and greenhouses and a well-stocked shop, but you also find yourself in a most welcoming environment, with a friendly and well-informed staff eager to chat and assist you. In fact, this is a good place to look around at leisure and learn about the plants that interest you—edible or not.

Gilbertie's Herb Gardens, Westport

Gilbertie's claims to be the largest herb plant grower in the country, with more than four hundred varieties of fresh herb plants to its name. (It supplies many nurseries and garden centers throughout the country.) Most of the herbs are actually grown on its large (thirty-seven acres) wholesale farm in Easton (unfortunately not open to the public), where twenty-three acres are dedicated to such herbs as artemisia, yarrow, sweet annie, sages and mints (of which there are many different kinds), anise, hyssop, and lamb's ear. The list is very long, and many of the names may not be familiar to most people. Among the specialties are Aussie sweet basil, coriander delfino, BBQ rosemary, lime basil, white flowering borage, lemon catnip, apple eucalyptus, silver leaf lavender, East Indian lemon grass, hot and spicy oregano, Cleveland sage, stevia, Nazareth sage, and silver lemon thyme. But even here in Westport, at this smaller retail facility, you can still get a good idea of what the farm offers, and you can buy or order any of its many varieties. Or you can simply enjoy looking at these specimens as you walk about the grounds, taking in your colorful surroundings.

A family enterprise since its beginnings in 1922 (it started as a cut-flower business), the farm is still run by the Gilberties. The present owner, Sal Gilbertie, is also coauthor of a new book on gardening, which offers practical—if somewhat unconventional—tips, as well as a good dose of garden philosophy. In his words, "When you are in the garden, you are closer to creation"; the gardener should "build good compost like they build good lasagna" (to encourage your plants to make friends); and "I like to see vegetable gardeners include some cutting flowers in their gardens." (Apparently certain flowers and vegetables like to live in symbiosis. Several other gardens in this book also feature combinations of vegetables and flowers growing nearby for mutual benefit.) The loving care given to the plants here is obvious; the gardens with their raised beds (to keep out pests) are neat and well tended.

You'll also discover unexpected touches here and there; one that we found particularly charming is a little wooden pigeon house, built to accommodate the many pigeons that have multiplied over the years and that seem to enjoy this home. And why would they leave? Gilbertie's is a happy place for all: plants, animals, and humans.

2·
ALL ABOUT BEES

An Intimate Look at Honey Making at
Red Bee Apiary, Weston

✒ DIRECTIONS

77 Lyons Plain Road, Weston, CT. Check the Web site for directions.

✒ INFORMATION

Open year-round. Before visiting you must call or e-mail to make an appointment. The best time to visit is from May to October, when the honeybees are at work; but you are welcome to come at other times, too, whether to tour the apiary or to purchase honey or skin-care or other products for sale. Telephone: 877-382-4618. E-mail: redbee@optonline.net. Web site: www.redbee.com.

✒ If you've ever been curious about bees and their crucial role in food production around the world, this is your chance to learn about one of nature's most fascinating processes at first hand: the making of honey. Although there are a number of apiaries in the region, the great majority of them do not allow visitors to walk around and see the hives up close. Fortunately, we found one that does—under careful supervision, of course. Red Bee is an artisanal apiary, owned and run by Marina Marchese, an expert on the fascinating subject of bees (as well as a designer and author), who seems completely dedicated to her work as beekeeper. She welcomes visitors (by appointment only) and guides them around the property, pointing out things of particular interest with great enthusiasm. You can actually see the bees at work (from late spring to early fall), examine the equipment in order to understand the process, ask questions, and sample the various types of tasty honey made here or in the vicinity (one of the highlights of the visit!).

The history of beekeeping goes back to the earliest civilizations. The ancient Egyptians, possibly the first professional beekeepers, used honey as a natural sweetener and to embalm their dead. For them bees had powerful symbolism. Chinese dynasties kept bees for medical practices and for the culinary pleasures of honey. The Greeks also cherished their honey (they still do, as evident in their yogurts and desserts), and the Romans sometimes used it as a bartering tool —and to pay their taxes.

Beekeeping is still very much alive today throughout the world (despite a recent nationwide epidemic), and apiaries in this region seem to be thriving, including Red Bee. One of several apiaries in the area, it is picturesquely set on a

Bees and Honey

And now a word about how honey is produced. Whoever coined the phrase "busy as a bee" knew a thing or two, because bees—that is, "worker" bees—are always busy. These bees (unlike the "drones," which do nothing but mate with the queen, which in turn only lays eggs and is fed royal jelly and kept in style) start to work from the time they are only a few weeks old and never stop until they die. The steps they take to make honey are surprisingly sophisticated. First, they gather up nectar by visiting hundreds of flowers in the vicinity (usually within two or three miles of the hive), suck it up, and store it in a special part of their stomachs. Once they are back at the hive, other bees mix the nectar with their own enzyme, producing a digestible substance. They then put the nectar mixture into honeycomb cells and with their fluttering wings wring out the superfluous moisture from this about-to-become honey. When the cells within a frame have been completely filled with honey, they are capped with beeswax (a substance secreted from the glands of worker bees).

By September the worker bees have usually finished making honey, and it's then time for it to be harvested. Beekeepers extract the honey from the "shallows," the shallow box at the top of the hive, which contains excess honey. This honey is then ready to be processed. (The other remains in the hive for the bees to eat.)

winding country road, typical of this rural slice of Connecticut. The leafy two-acre property includes a little red house (where Marchese lives and works), vegetable and flower gardens to attract the bees, a chicken coop, and, of course, the hives—set in a sunny and protected spot in back, against a hillside. (The site adjoins a parkland, which gives the bees even more opportunities to find nectar.)

Red Bee Apiary has eleven hives. There are traditional Langstroth hives (named after a noted nineteenth-century beekeeper whose design modernized the process), and one is a top bar hive. The former consist of stacked rectangular wooden boxes and look something like filing cabinets. The design has been carefully planned to make the life of the bee colony as stressfree as possible —and the work of the beekeeper easier. The main part of the hive includes two boxes that contain removable wooden frames lined up in rows. Here the honeybees build their comb and raise their brood. The third box on top of the hive body is the one that contains the excess honey, which is then harvested. Other parts of the hive are an outer cover to protect the hive from inclement weather, an inner cover for further insulation, a bottom board where the bees can enter the hive, a feeder box filled with sugar syrup, and a hive stand at the base.

The inside of the house, where the final honey production takes place at Red Bee, is filled with various pieces of equipment and other items relating to honey making. A stainless-steel circular tank called a "spinner extractor" is used to place the frames of the "shallow" box to extract the remaining honey. The honey then passes through a "honey gate" and is strained into a bucket. It is now ready to be eaten.

We spent much of our time in the kitchen, where the honey is strained and tastings take place. (Marchese hosts tastings for the public on a regular basis—something not to miss!) Sampling a wonderful assortment of honey produced here (and at nearby apiaries), including blueberry, alfalfa, and buckwheat honey, was a new and wonderful experience for us. (Red Bee makes a delicious dark wildflower honey, as well as honeycomb and creamy honey.) Finally, check out a nearby room featuring quite an assortment of items for sale made from beeswax: beauty products, therapeutic creams, and candles, in addition to honey for eating, impossible to resist! Copies of Marchese's recent book, *Honeybee: Lessons from an Accidental Beekeeper*, are also on display. A comprehensive and very clear guide to beekeeping and everything relating to bees, the book will give you all the information you will ever need to know on the subject. We recommend it as a fascinating read to complete your visit at Red Bee.

3 •

WHERE RHUBARB AND WINEMAKING MEET IN THE LITCHFIELD HILLS

White Silo Farm and Winery, Sherman

🐾 DIRECTIONS

32 Route 37 East, Sherman, CT. Take I-84 to Danbury and follow signs to Route 37 north (exit 6). Follow Route 37 through New Fairfield and through the small town of Sherman. White Silo is just north of town on your right.

🐾 INFORMATION

The tasting room is open from April through December on Fridays and weekends from 11 A.M. to 6 P.M., but you can visit and watch the making of wine, as well as attend a great variety of events held here, by phoning for information. The owners are remarkably hospitable to visitors. Telephone: 860-355-0271. Web site: www .whitesilowinery.com.

🐾 We had no idea when we visited here what a fruit winery would be like. This one is a delight to visit, and the owners are so hospitable that we highly recommend this outing. Here you can actually see such a variety of fruits being

Rhubarb at White Silo Farm and Winery, Sherman

grown and turned into wine (and you can taste them all) that you will be surprised; you may never have thought of blackberry wine or black currant wine or raspberry wine—not to mention rhubarb wine.

This is an unpretentious setting: a hilly area of the farm covered with rhubarb's giant leaves, a field below of blackberries, and so on, all marked by the eponymous big white silo and a comfortable nineteenth-century former dairy barn where the winemaking takes place and where you can sit and try each and every type of wine with cheese and crackers. The fruits are grown on four acres of the twenty-acre farm. You can also sit outdoors in the bucolic setting overlooking the farm, sipping White Silo's blackberry sangria or comparing currant wine to raspberry wine to rhubarb wine.

This is not the type of vineyard visit that features an expansive, lookalike Tuscan villa or a giant party venue. It is homespun and pleasant, with monthly local art shows, old rafters, and a free-and-easy atmosphere. "We have worked very hard to create a unique experience at our farm and winery," comments the owner.

We had a tour of the produce that is grown here (without any pesticides) and the barn facilities, with their seemingly high-tech vats and machinery; there are fermentation, bottling, and corking rooms. In fact, White Silo turns out some ten thousand bottles of wine (and cassis) per year. (All are available to buy at the winery.) It gets many awards.

And if you come at the right time of year, you can pick from the vast crop of raspberries to take home with you.

4 ·
ENJOYING A NUT WALK
A Stroll through Bartlett Arboretum, Stamford

❧ DIRECTIONS

151 Brookdale Road, Stamford, CT. Take the Merritt Parkway to exit 35 and head north on High Ridge Road (Route 137). Go one mile to Brookdale Road, turn left, and you'll find the entrance on your right. The Arboretum is about six miles north of Stamford.

❧ INFORMATION

Open daily, 8:30 A.M.–sunset (visitors' center until 4:30 P.M.). There are guided tours available. All trees are identified. Admission is free. Telephone: 203-322-6971.

❧ This beautiful sixty-three-acre arboretum has wonderful trails, woodland settings, wetlands (with an elevated walkway), and more. Part of the University of Connecticut's Department of Plant Science, it is a great place to wander around—and if you are a nut enthusiast, it is a particularly interesting place. For in these acres there are numerous nut trees: pecans, heartnuts, walnuts, and chestnuts among them. (Since commercial nut farms are uncommon in this area, we are lucky to find this large a collection available to us.)

These trees are scattered through the woodland areas, and they drop their nuts around them. (You are welcome to pick them up and take them with you.) We recommend calling before you go, to discover just when the nuts will fall, and going to the visitors' center to inquire where to look for your favorites.

While you are at the arboretum, you might also visit the gardens and special collections of evergreens, rhododendrons, azaleas, and so much more. But pay particular attention to the nut trees, as nuts play a central role in many diets and cuisines and are certainly a delicious part of nature's bounty.

5·

MAPLE SUGAR TIME

A Late-Winter Treat at McLaughlin Vineyards, Sandy Hook; Flanders Nature Center, Woodbury; and Stamford Museum and Nature Center, Stamford

✍ DIRECTIONS

McLaughlin Vineyards: 14 Albert's Hill Road, Sandy Hook, CT. Take I-84 east from Danbury to exit 10. Turn right at the end of the ramp onto Route 34, go through a traffic light, and take the first left onto Walnut Tree Hill Road; go about two miles to a grass island. Bear left onto Albert's Hill Road; the entrance is on your right.

Flanders Nature Center: 5 Church Hill Road, Woodbury, CT. Take I-84 to exit 15, to Route 6 east. Turn left on Route 47 at Hotchkissville and continue onto Route 132. Turn right on Brushy Hill Road, continue to Flanders Road, and take a left to the Nature Center at the intersection of Church Hill Road.

Stamford Museum and Nature Center: 39 Scofieldtown Road, Stamford, CT. Take the Merritt Parkway to exit 35, to Route 137 north. Follow signs.

✍ INFORMATION

For general information contact the Maple Syrup Producers Association of Connecticut, at 860-688-1718 or www.ctmaple.org.

McLaughlin Vineyards: Telephone: 860-599-9463. Web site: www.mclaughlinvineyards .com.

Flanders Nature Center: Telephone: 203-263-3711. Web site: www.flandersnaturecenter .org.

Stamford Museum and Nature Center: Telephone: 203-322-1646. Web site: www .stamfordmuseum.org.

✍ For a few late-winter weeks you can watch maple sugar being tapped and made (and taste it!) in several spots in nearby Connecticut. This centuries-old, venerable tradition is alive and well wherever there are sugar maples or black maples, and, we discovered, there are many thriving maple-sugaring enterprises in the state that invite visitors to watch the proceedings, to enjoy tasting events, and to bask in the heady scent of this New England favorite. (Many rural families also tap their own maples.)

The origins of maple syrup go back to Native Americans; they taught the colonial settlers the simple process, though it was the settlers who devised the small taps that were similar to those that are still used today.

Because Connecticut is a bit warmer than the usual maple syrup locations of Vermont, New Hampshire, and Maine, the season is a bit earlier: generally mid-February through most of March, depending on the weather. (We recommend calling a couple of weeks in advance, perhaps in late January, to find the right time to visit.) The daytime weather should be in the midforties for several consecutive days, with nighttime temperatures still below freezing, for the sap to travel from the roots of the tree up the trunk of the tree to the branches and the buds. This is when tapping occurs.

When you visit, you'll see the harvesting of the sap, which is done by drilling a tiny hole slightly upward through the bark and using a tap connected to a bucket or a tube. Buckets are collected twice a day. Sap is brought back to a small building known as a sugarhouse, where the simple process of evaporating and boiling into the delicious sweet syrup we all know and love takes place. Already clear when it is tapped, the syrup is fed into an evaporator in the sugarhouse and boiled for several hours (reducing water in the sap), making it thicker and darker. One gallon of maple syrup may require thirty to forty gallons of maple sap. The sap is then filtered and bottled. The lightest syrup (used primarily for maple sugar candy) is made early in the season, while the "dark amber" —the general favorite—is made later. All of this is done before one's very eyes.

This is definitely an outing that children will enjoy. To visit a maple-sugaring enterprise with children, you might want to go to one of the nature centers in the following list, as they have educational programs explaining the procedure (and even allowing the kids to "help"). Adults will also enjoy any of the following sites for their rustic charm. (And you can buy the syrup too!) Among our favorite maple-sugaring sites are these:

McLaughlin Vineyards in Sandy Hook: This is a small part of the elegant vineyards (see "Two Delightful Vineyard Visits"). Here you walk across a field surrounded with grapevines and lovely scenery to find a small sugarhouse bordered by dozens of maple trees, with a plume of delicious-smelling smoke rising from the chimney. (This is definitely a pleasant place to visit, but of course you can't combine it with your vineyard visit because of the season.)

Flanders Nature Center in Woodbury: This is an excellent educational spot, which includes a pretty hike through the maple woods to the sugarhouse, lectures, educational events, and tastings. You can also spend time here hiking on a number of trails (Geology Trail, Wilderness Trail, etc.) and in a variety of classes.

Stamford Museum and Nature Center in Stamford: Here you walk through a small farm, with chickens and cows and barns, as well as the little sugarhouse. Among the attractions is an exhibition of how maple syrup was made in the good old days.

6·

A PICTURE-PERFECT CONNECTICUT FARM FROM THE 1840s

"Celebrating the Cycle of the New England Year" at Warrup's Farm, Redding

✍ DIRECTIONS

51 John Read Road, Redding, CT. Take the Merritt Parkway to exit 39B, to Route 7 north, to Route 107 east, to Redding Center. From the center of town continue a mile and a quarter to John Read Road, on your right, opposite the country club. Take that road, and you'll find the farm on your left (a large white house).

✍ INFORMATION

We suggest you call before visiting; the farm has many events (such as pumpkin picking) that would be an added treat. Telephone: 207-938-9407. Web site: www.warrups farm.com.

✍ We have visited all kinds of farms as we have driven through this area, but none is more picturesque—or more inviting—than Warrup's Farm. A family-owned farm since the 1840s, it features a magnificent mid-nineteenth-century white house, rolling fields, thriving vegetable beds, and unusually hospitable owners. In addition, there are occasional you-pick-it days, and the owners "invite the public to participate in farming," with programs for kids and interns.

This is an unusually large farm by Connecticut standards: some three hundred acres. Unlike the great farms of the Midwest, however, these are not flat

---

Why Is Connecticut called the "Nutmeg State"?

Connecticut, officially known as the "Constitution State," has a second—unofficial—name: the "Nutmeg State." Some people say that the state got this nickname from its sailors returning from afar carrying nutmeg, which was a valuable spice in New England during the eighteenth and nineteenth centuries. Others claim that the name came from Connecticut peddlers who sold small carved nutmeg-shaped knobs to local Native Americans and others—for whatever reason. There may be other equally colorful stories, as well. It's a name that raises questions, since the nutmeg—a hard, aromatic kernel of the seed of a tree originally from India—is hardly native to Connecticut.

acres but encompass a variety of gently rolling fields, wooded areas, and wetlands along a pretty, old-fashioned road. Owned and operated by the Hill family since the 1840s, Warrup's Farm began its present operations in the 1970s: growing vegetables, herbs, Christmas trees, and flowers; producing maple syrup; and making hay.

About fifty years ago the family bequeathed the farm to the Redding Land Trust, securing the land forever for agricultural uses (becoming a model for other landowners in a crowded state). The owners take their mission of sustainable farming and responsible use of the land seriously; all produce is certified organic.

As the farm's Web site states, Warrup's celebrates the cycle of the New England year: late winter is maple syrup time. The first three weekends in March you can visit the sugarhouse and watch as the sap is boiled into syrup. In summer you will see salad greens harvested in early July, with vegetables, herbs, and flowers at their peak in August. (Pick-your-own available.) In October pumpkins are ripe, as are Indian corn, gourds, and other fall delights. (Hayrides are available too.) In December Christmas trees are ready for cutting.

7·
TWO DELIGHTFUL VINEYARD VISITS

Walking and Tasting at McLaughlin Vineyards,
Sandy Hook; and Hopkins Vineyard, Warren

✣ DIRECTIONS

McLaughlin Vineyards: *14 Albert's Hill Road, Sandy Hook, CT. Take I-84 east from Danbury to exit 10. Turn right at the end of the ramp onto Route 34, go through a traffic light, and take the first left onto Walnut Tree Hill Road. Go about two miles to a grass island. Bear left onto Albert's Hill Road; the entrance is on your right.*

Hopkins Vineyard: *25 Hopkins Road, Warren, CT. Take I-84 east to exit 7, to Route 7 north and Route 202 north. Then take Route 45 north at New Preston. Continue on Route 45 to North Shore Road (on left), to Hopkins Road (second right).*

✣ INFORMATION

McLaughlin Vineyards: Open November–May, Wednesday–Sunday, 11 A.M.–5 P.M.; June–October, daily, 11 A.M.–5 P.M. Closed on major holidays. Telephone: 203-426-1533. Web site: www.mclaughlinvineyards.com.

Hopkins Vineyard: Open May–December, daily; January–February, Friday–Sunday; March–April, Wednesday–Sunday. Hours are 10 A.M.–5 P.M. throughout the year. Telephone: 860-868-7954. Web site: www.hopkinsvineyard.com.

✣ Of Connecticut's expanding number of vineyards, two of our favorites are McLaughlin Vineyards and Hopkins Vineyard. Both are situated in the western part of the state; as they are not far from each other, they can be easily visited in one day's outing or, if you prefer, on separate occasions. Both offer everything you could wish for in a vineyard: a welcoming atmosphere, a beautiful setting with inviting walks, and of course fine wines to be tasted and enjoyed!

McLaughlin Vineyards and Winery, set on 160 bucolic acres of meadows and woodlands along the Housatonic River, is truly off the beaten path. Surrounded by a state forest and nature conservancy, it takes some driving on winding roads to reach this secluded spot. But it's well worth the effort, as you are rewarded with an exceptionally lovely site to visit.

This is a family owned and operated vineyard/winery, with its own rustic, yet elegant, personal style. On a sunny hillside you'll find the fifteen-acre vineyard, where you are welcome to stroll and enjoy breathtaking views. (Occasional tractor rides in the vineyard are offered.) Nearby is a converted old barn now used as a winery and tasting room, with outdoor seating overlooking flower gardens. Beyond are deep woods with tall sugar maple trees and a sugar "shack"

Vineyard in spring (iStockphoto.com/Susan Trigg)

for making maple syrup. (Maple-sugaring demonstrations have become quite popular in this region. McLaughlin's are especially hands-on, and you can tap the maple trees, collect the buckets filled with sap, and then watch it become maple sugar.)

You'll want to sample the artisanal wines produced here, which come only from estate-grown or other local grapes. (Tastings are offered during normal business hours, for a small fee.) The whites, made from hybrid varietals and vinifera, include a nice chardonnay, a semisweet Coyote made from estate grapes, and a sweet Snow Goose. The red wines include a refreshing rosé and a delicious merlot.

If you are a hiker and nature lover—in addition to a wine fancier—you will especially enjoy being here. Included within the property is a fifty-acre wildlife preserve and bald eagle sanctuary. From mid-December until the end of March, when these birds migrate south from Canada, you might spot them flying over the vineyard. Or join an organized eagle viewing, followed by wine tasting at the winery (call 800-368-8954). Hikers have many (mostly) well marked and maintained trails to choose from within the nearby state forest. (Ask for directions at the winery.) Picnicking is allowed on the grounds.

Hopkins Vineyard is an equally spectacular site, but in a very different way. Situated on a hillside, high above Lake Waramaug, the thirty-acre vineyard and winery enjoys unimpeded views of the lake and surrounding countryside, giving it a feel of open space. As welcoming and friendly in ambiance as McLaughlin Vineyards, Hopkins, too, offers many amenities: wine tastings (also for a small fee) in a converted two-hundred-year-old barn, guided tours on weekends, and special events on the third weekend in May and September—not to mention a spectacular drive around the lake, which you do on your way to the vineyard and back. But we assume you'll also want to explore the grounds on your own and to savor the surroundings at your own pace.

We are told that this is the only vineyard in the state with its own microclimate (influenced by the nearby lake), resulting in a longer growing season. The vines themselves are perched on a hill facing south, which also works as a tempering factor, weatherwise. As you walk around, look for several varieties of vinifera and French hybrid grapes (or ask someone to identify them for you). The wines produced here (many of which, except for the sparkling wines, are available for tastings) are classical grape varieties, including chardonnay, cabernet, franc, and merlot. Right next door to the vineyard is the elegant

Hopkins Vineyard, Warren

Hopkins Inn, where you can also enjoy some of these wines, served along with excellent food.

Hopkins Vineyard sits on a historic property, once known as Hopkins Farm. Dating from 1787, for years it produced tobacco and raised sheep and horses. In 1979 the dairy farm became a winery and opened to the public shortly thereafter. Since then, it has been one of the region's main attractions—due to its wonderful location, as well as its highly regarded wines and many special events.

8 ·

FROM BRAZILIAN VEGETABLES TO APPLES AND PUMPKINS

A Day at Holbrook Farm and Blue Jay Orchards in a Rural Connecticut Community, Bethel

❧ DIRECTIONS

Holbrook Farm: *45 Turkey Plain Road (Route 53), Bethel, CT. Take I-84 to exit 5 to Danbury. Follow signs through Danbury; the road you are on becomes Route 53. At the War Memorial follow Route 53 to your left into the town of Bethel. From the intersection of Routes 53 and 302, go two more miles on Route 53; the farm is on your right.*

Blue Jay Orchards: *125 Plumtrees Road, Bethel, CT. Take I-84 to exit 8. Take Route 6 east for a mile and three-quarters to Hawleyville Road; go a third of a mile to the second stop sign and turn right onto Plumtrees Road.*

❧ INFORMATION

Holbrook Farm: Open to visitors in season, daily, 9 A.M.–6 P.M. Telephone: 203-792-0561. Web site: www.holbrookfarm.net.

Blue Jay Orchards: Open end of August–December, 10 A.M.–5 P.M. Telephone: 203-748-0119. Web site: www.bluejayorchardsct.com.

❧ There are several unusual features to this farm/orchard outing in the rural community of Bethel. Nearby Danbury is home to a large community of Brazilians, and Holbrook Farm offers them and us true Brazilian vegetables, as well as more familiar crops. The farm began planting Brazilian foods when the owners were brought some seeds and asked if they could supply the community with vegetables. You are welcome to walk around the thirteen acres in season and see (or purchase at the shop) such unexpected vegetables as a small, delicious, spiny cucumber called maxixe; jilo, which resembles our eggplant (it is grown in the greenhouse in the winter); quaibo; collard greens; okra; and couve. (Look for the Brazilian flag hanging out front when the farm has these vegetables.) It produces free-range chickens and fresh eggs. No pesticides or herbicides are used.

Not too far away is one of the nicer apple orchards in this part of the world. Blue Jay Orchards is a 140-acre farm, one of the few remaining orchards left in Fairfield County. It was the first farm to be preserved as farmland by the state of Connecticut and, as such, was purchased in 1985 by the current owners. They invite pick-your-own visitors, who can number as many as ten thousand people during the autumn picking season.

What you will find when you visit is a spacious orchard and farm with several varieties of apples and pumpkins (ready to pick in October), as well as a cider mill, where you can see (and taste) the orchard's own apple cider.

9.

TWO NEIGHBORING VISITS AMID FLOWERS, VEGETABLES, AND VINES

White Flower Farm and Haight-Brown Vineyard
Invite You to Walk Around, Litchfield

᨞ DIRECTIONS

White Flower Farm: *167 Route 63, Litchfield, CT. Take I-84 east to exit 17. Take Route 63 north to Watertown. Continue on Route 63 north for about thirteen miles. The farm will be on your left. It is well marked; look for elegant displays by the road.*

Haight-Brown Vineyard: *29 Chestnut Hill Road, Litchfield, CT. From the center of Litchfield, follow signs for Routes 202 and 118, veering to the right (Route 118) at the intersection. Take Route 118 (East Street) to Chestnut Hill Road (a right turn). The vineyard is up ahead and is clearly marked.*

᨞ INFORMATION

White Flower Farm: Open March–late October, daily, 9 A.M.–5:30 P.M. Telephone: 860-496-9624. Web site: www.whiteflowerfarm.com.

Haight-Brown Vineyard: Open May–November, Monday–Saturday, 11 A.M.–5 P.M.; and Sunday, noon–5 P.M. Open only on weekends during winter months. Tours by appointment only. Telephone: 860-567-4045. Web site: www.haightvineyards.com.

᨞ Litchfield, that delightful New England village which is on everybody's list of places to visit in the region, is also very close to at least a couple of nature sites worth seeing. White Flower Farm, just a few miles south of the center of town, is a remarkable nursery in a class of its own. Looking more like an impeccably maintained botanical garden or English garden in a private estate than a commercial enterprise, it embodies nature's bounty at its very best. The lush grounds include five acres of display gardens and elegant greenhouses filled with flowers and vegetables, all within a vast property of pasture, forest, and wetlands. (In fact, the bucolic feel to this lovely place is enhanced by the presence of cattle and sheep grazing on fields just across the road from the nursery.)

Since 1950, when White Flower Farm was founded, it has focused on "finding, testing, growing, and delivering the very best ornamental and edible plants to fellow gardeners throughout the country," as its brochure indicates. To achieve this goal, members of the friendly and helpful staff are always available to answer questions and offer suggestions. Various events, lectures, and demonstrations are given to broaden gardening techniques and knowledge. An annual spring event call "Tomatomania" celebrates the versatile tomato in all its aspects. Included are colorful displays, a sale of over one hundred varieties of tomato seedlings (including staff favorites such as Sungold, Taxi, Carmello, Jaune Flammee, Strawberry Oxheart, and Cherokee Purple), and a variety of talks on subjects relating to this favorite American staple. (A recent talk offered tips on how to plant tomatoes and other edibles on decks and patios.)

From the time you arrive at the farm you will want to explore these grassy acres—and you are invited to do just that (and without pressure to buy anything at all). A self-guided walking tour is available, and you can wander about on you own, viewing gardens of various types—decorative, demonstration, what have you—while perhaps planning your own garden or just gathering information. You'll notice a very long border (about 290 feet) typical of English gardens. Designed by Fergus Garrett, an associate of the late Christopher Lloyd of Sussex (England), this arrangement of multicolored flowers is truly a delight. Visit the spacious and airy greenhouses to view smaller decorative plants, as well as the many edibles grown here: tomatoes, peppers, zucchini, onions, lettuces, herbs, and more. We are told that planting vegetables is a good way to start as a gardener, as it teaches you about such basics as climate, the life cycle of plants, and their anatomy—not to mention the pleasure of eating what you have grown!

Before leaving the nursery make sure you stop at the elegant shop to pick up the farm's comprehensive catalog. This thick volume includes valuable information and listings of all the plants available for shipping—over three hundred of them. (If you weren't a gardener before visiting White Flower Farm, you will surely be inspired to become one after!)

To reach our second suggested site just outside of Litchfield, you need to drive back through town, following the directions given at the beginning of the chapter. Haight-Brown Vineyard is a family-run vineyard and winery located on a serene, fairly flat, ten-acre property, above the Bantam River. The terrain is fertile, ideal for producing wines including Riesling and chardonnay. The winery also has the distinction of being the first in the state, having opened in 1978, just after the Connecticut legislature passed a Farm Winery Act.

You'll enjoy a visit at this fairly low-key winery/vineyard. Unlike most we have visited, this one actually features a "vineyard walk," where you are encouraged to amble freely on your own through rows of vines. Or you might prefer taking a guided tour of the vineyard (available by appointment only). After exploring the outside, find the indoor wine bar, choose your wines for tasting, and walk over to the terrace to enjoy the view while you sip. (In winter you can sample wine around a cozy fireplace.)

During the year various events are held, so that you can learn more about wines and cheeses and related subjects. In the fall an annual Harvest Festival takes place, with hayrides, grape stomping, locally brewed beer, and other seasonal pleasures—in addition to wine tasting, of course.

10·
THE THREE EXTRAORDINARY
HERB GARDENS OF COVENTRY

Exploring Edmondson's Farm, Caprilands Institute, and Topmost Herb Farm, Coventry

⚘ DIRECTIONS

Edmondson's Farm: *2627 Route 44 (also called Boston Turnpike), Coventry, CT. Take I-84 east to exit 59, I-384. From the light at the end of I-384 continue two and a quarter miles on Route 44. Located near the town line of Bolton.*

Caprilands Institute: *5334 Silver Street, Coventry, CT. Take I-84 east to exit 59, I-384. From the light at the end of I-384 continue onto Route 44 east. Silver Street is a right-hand exit from Route 44.*

Topmost Herb Farm: *244 North School Road, Coventry, CT. Take I-84 to exit 67, Route 31 south. Continue for three and a half miles, then take a left onto North School Road. Farm is the fourth driveway on the left.*

⚘ INFORMATION

Edmondson's Farm: Open for you-pick-it fruits as well as visiting its herb gardens. Telephone: 860-742-6124. Web site: www.edmondsonsfarm.com.

Caprilands Institute: Open daily except major holidays. We recommend a visit in spring or summer. (In the winter it is open only in the afternoons.) There are numerous special events, including lectures and demonstrations of herbal uses, and tea is served too.

Topmost Herb Farm: Telephone: 860-742-8239. Web site: www.topmostherbfarm.com.

⚘ We don't know precisely why Coventry is especially hospitable to herbs, but you'll find three interesting herb farms in this one small place. (Although it is slightly farther than our usual tristate visits, we thought it was worth an extra half hour if you are an herb enthusiast.)

The first venue is Edmondson's Farm, a thriving business with about two hundred kinds of herbs. You'll enjoy seeing them growing here, and you can purchase them (and a variety of other produce) in the farm's shop. You can contact the farm in advance of your visit to learn what herbs and fruits are ready and can be picked or purchased, or you can just enjoy the wonderful mix of nature's bounty.

Caprilands Herb Farm, now called the Caprilands Institute, was the creation of the redoubtable Adelma Simmons, who for more than sixty years guided its unusual path. On what was once a rock-strewn Connecticut dairy farm (bought by her family in 1929), Simmons built an irregular series of separate gardens with whimsical names and intriguing quotations propped among the flowers. Typical low stone walls divide the fifty-acre landscape. Primarily an herb farm growing hundreds of varieties of herbs, Caprilands also includes any number of flowers and shrubs (and weeds too). There is nothing pristine or formal about this landscape—in fact, that is part of its great charm.

Each separate bed has a name and a theme illustrated by the flowers or herbs chosen by Simmons. She used to spend each long day, except for a few holidays, overseeing the gardening and regally receiving curious and admiring visitors from her chair in the shop in the weathered barn (formerly the milk house). There you can purchase herbs and her books—some fifty different titles, including her best-known *Herb Gardening in Five Seasons*.

Among the thirty thematic gardens that caught our attention were the Medieval Garden (in which all the flowers and leaves are silver), the Garden of the Stars (in which plantings are divided into twelve culinary beds based on signs of the zodiac), the Saints' Garden (adorned with small statuary and planted with symbolic plants such as rosemary and Madonna lilies), the Shakespeare Garden (with appropriate quotations), the Bride's Garden (in which two hearts outlined in brick are filled with symbolic plants representing love throughout history—lemon verbena, forget-me-nots, orange trees), and more practical gardens such as the Cook's Garden and the Salad Garden. There are plots devoted to colors—gold, blue, white—and there are gardens for dyeing colors, for potpourri, for flowers that are good for drying or for fragrance, and even for onions.

Most of these small, patterned beds are delightful in their own miniature way. (For the serious visitor, a guidebook by Adelma Simmons—who was exceptionally knowledgeable about the history and herbal uses for each plant—is available. It identifies the plants and describes the symbolism, preservation, and practical uses of each.) For those who prefer literary references, handwritten quotations abound.

About tansy (*Tanacetum*): "On Easter Sunday be the pudding seen / To which tansy lends her sober green" (from *The Oxford Sausage*).

About onions: "This is every cook's opinion, / no savory dish without an onion, / But lest your kissing should be spoiled / Your onions must be fully boiled" (Jonathan Swift).

About broom: "I'm sent with Broom before / To sweep the dust behind the door" (Shakespeare, from *A Midsummer Night's Dream*).

At Topmost Herb Farm forty varieties of heirloom tomatoes add to the herb mix. You can see all of the herbs—and vegetables (grown without pesticides, using an organic potting mix including seaweed and kelp)—and can purchase them as plants or produce. Among the farm's specialties are medicinal and culinary herbs on display. Gardeners, don't miss the garlic and sweet onions.

If you visit Coventry on a Sunday from June through September, you'll find the Coventry Regional Farmers' Market in town.

11 ·
A POTPOURRI OF VEGETABLES, FLOWERS, AND OUTDOOR SCULPTURE AMID THE GREENERY

The Picturesque Garden of Ideas, Ridgefield

✿ DIRECTIONS

647 North Salem Road, Ridgefield, CT. Take I-684 north to exit 7 (Purdys/Somers), and take Route 116 east toward Ridgefield. The garden is located on Route 116 (North Salem Road), about a mile and a third from the state line and about three miles northwest of the center of Ridgefield. Look for a small green-and-white sign on the right and a rustic fence.

✿ INFORMATION

Open mid-March–end of November, Monday–Sunday, 8 A.M.–6 P.M. (open until 7 P.M. during summer months). Telephone: 203-431-9914. Web site: www.gardenofideas.com.

✿ You could call the Garden of Ideas a boutique farm, an imaginative garden, or an intimate outdoor sculpture gallery: it's all of the above. Describing itself in its brochure as "the destination of choice for plant enthusiasts, nature lovers, art collectors, tomato growers, slow foodies, and bird freaks," we found it to be one of the most creatively designed green spaces we have discovered in our explorations.

Set on acres of marshes, woodlands, and meadows, this bucolic site combines native plants artfully set amid the greenery. Here herbs and peppers live alongside unusual woody plants and herbaceous perennials in a natural-looking —yet well-tended—arboretum-like setting. It's fun to explore the grounds, as you make your way along winding paths through little gardens and farm plots, noting the trees and decorative sculpture (which, by the way, is for sale and changes regularly). There is a great deal of whimsy in the way everything has been put together, something that will amuse children, too (so be sure to bring them along).

In the early twentieth century the property was part of a large family farm and gristmill. When farming stopped by the 1950s, a few houses were built, including the owners' present house. The six-acre garden was created in 1995 by Joseph Keller (a family member) and Ilsa Svendsen, who runs it. (The remaining portion of the twelve-acre property, including the lovely old barn, which was part of the original farm, is still privately owned.)

After you have taken your walk, be sure to check out the farm stand, located right on the premises; here, depending on the season, you will find a wide variety of sustainably grown local vegetables—including the more exotic Asian greens—as well as herbs, vegetables plants, and seeds. Locally grown honey and other delicacies are often available, as well. If you enjoyed the outdoor artwork, stop by the small indoor gallery, where you'll also find wall murals and various artifacts for sale. We hope that a visit to the Garden of Ideas will give you a few ideas for your own garden at home!

Garden of Ideas, Ridgefield

AND KEEP IN MIND . . .

12 • The Hickories, Ridgefield

ADDRESS: *126 Lounsbury Road, Ridgefield, CT.*
INFORMATION: Telephone: 203-894-1851. Web site: www.thehickories.org.

This is an old farm that has been lovingly revived by the new generation of the family that owned it from 1936 to 1991. Today this rustic spot with traditional Connecticut stone walls uses permaculture, growing mostly perennials (such as asparagus, garlic, currants, grapes, and berries) with a holistic approach. The young farmer has made part of the farm a CSA (community-supported agriculture), inviting other farmers to share in the land. This is an ongoing enterprise, with new necessities every year: one of the latest is very high deer fencing. There are tours and you-pick-it events at this scenic farm. Contact before visiting.

13 • Jones Winery, Shelton

ADDRESS: *606 Walnut Tree Hill Road, Shelton, CT.*
INFORMATION: Telephone: 203-929-8425. Web site: www.jonesfamilyfarms.com.

This winery is part of a family farm; it includes twenty-five acres of berry crops used to make fruit wines, as well as traditional grapevines on seven additional acres. The tasting room is housed in a nineteenth-century dairy barn; six generations of the family have farmed here.

Why Join a CSA?

1. CSAs follow sustainable farming practices, which are better for both the environment and the economy.
2. You have access to the freshest, highest quality organic (usually) produce.
3. The farmer has a guaranteed market for all produce.
4. CSA farming can be a wonderful learning experience for the entire family (kids included).
5. CSAs provide greater diversity in food choices.

14 • Jerram Winery, New Hartford

ADDRESS: *535 Town Hill Road, New Hartford, CT.*
INFORMATION: Telephone: 860-379-8749. Web site: www.jerramwinery.com.

Unusually picturesque, this winery in a historic town is set on a hill (one thousand feet elevation), enjoying a long growing season in the sun. It makes different wines and offers tours, picnicking, and tasting, among other treats.

15 • Land of Nod, East Canaan

ADDRESS: *99 Lower Road, East Canaan, CT*
INFORMATION: Telephone: 860-824-5225. Web site: www.landofnodwinery.com.

This nationally known winery has been in operation for nine generations. It is situated on two hundred hilly acres and offers a variety of fruit wines (raspberry and blueberry blends), maple sugar, and award-winning wines (chardonnay and pinot noir). The state considers it a "Farm of Distinction."

16 • Lyman Orchards, Middlefields

ADDRESS: *32 Reeds Gaps Road (junction of Routes 147 and 157), Middlefields, CT.*
INFORMATION: Open daily, 9 A.M.to 6 P.M. (September–October, 9 A.M. to 7 P.M.). Telephone: 860-349-1793. Web site: www.lymanorchards.com.

This one-thousand-acre farm, established in 1741, is one of the oldest family-owned businesses in the entire country. Its scenic hillside orchards include pick-your-own apples, peaches, pears, berries, and pumpkins; in the fall you'll find a sunflower maze and a corn maze, which are fun to explore. This family-friendly site also offers picnicking on the grounds and deck, horse-drawn wagon rides, and pony rides. You may be surprised to hear, too, that within the farm property itself are two eighteen-hole championship public golf courses—not something you expect to find on a farm!

NEW YORK CITY

17·
A CULTURAL GEM WITH BOTANICAL GARDENS

Pumpkins, Herbs, and Quince at Wave Hill, Bronx

❧ DIRECTIONS

675 West 252nd Street, Bronx, NY. From Manhattan take the West Side Highway up to Riverdale. After the Henry Hudson Bridge toll booths, take the 246th Street exit. Drive on the parallel road north to 252nd Street, where you turn left and go over the highway. Take a left and drive south on the parallel road, to 249th Street, and turn right. Wave Hill is straight down the hill. There is limited parking on the grounds and also street parking. There is a free shuttle van at the front gate to and from MetroNorth's Riverdale Station and the 242nd Street stop on the number 1 subway line.

❧ INFORMATION

Open year-round, Tuesday–Sunday, 9 A.M.–5:30 P.M. (closes at 4:30 P.M. from mid-October to mid-April). Conservatory hours: 10 A.M.–noon, 1 P.M.–4 P.M. Admission is free on Tuesday and Saturday mornings until noon. We recommend visiting during the week, when Wave Hill is not crowded. The park is a joy to visit at any time of the year, including winter. Telephone: 718-549-3200. Web site: www.wavehill.org.

❧ Wave Hill is one of New York City's less known gems. Although familiar to most gardeners, this rare botanical garden and art/environmental center comes as a real surprise to most first-time visitors. Its picturesque setting high above the Hudson River (with remarkable views), its vast rolling lawns dotted with huge old trees and occasional sculpture, its acres of woodlands, and especially its internationally acclaimed gardens—including a charming vegetable/herb garden with eclectic plantings—make this twenty-eight-acre park a unique spot. And as you stroll by its two stately manor houses set amid the plantings, you'll imagine you're enjoying a day at a private estate, miles away from the city.

In fact, in the past Wave Hill was the country home of several prominent New Yorkers. From the time the first of its two houses was built in 1848, it was occupied by illustrious people who often entertained members of New York society. Teddy Roosevelt spent a summer here as a boy (it is said that he learned to appreciate nature here); William Makepeace Thackeray visited on occasion; Mark Twain lived here from 1901 to 1903 (and even built a tree house on the grounds); and Arturo Toscanini occupied the house from 1942 to 1945.

Though Wave Hill (also called Wave Hill Center for Environment Studies) sponsors indoor and outdoor art shows, horticultural exhibits, chamber music

Wave Hill, Bronx

concerts, arts festivals, and more, it is above all a place in which to savor a group of extraordinary formal and wild gardens—including some with edible plants—both indoor and outdoor. Largely the creation of Marco Polo Stufano, the former director of horticulture, these gardens (six "garden rooms," pus a conservatory and greenhouses) are surprisingly intimate, in contrast to the grandeur of the surrounding landscape. The plantings have been designed on a small scale, separated by grassy areas and paths; the result is an inviting and personal environment, in keeping with Wave Hill's tradition as a private estate.

The diversity of plants is exceptional, including more that one thousand classifications of plants and well over three thousand species; and the garden design is unusually creative and unconventional, with color as one of the guiding principles. In addition to the flower gardens (both formal and less so) that you would expect to find on such an estate, a few contain edible plants, too. Two of these, the Dry Garden and the Herb Garden, stand out. Built on the stone foundation of a former greenhouse, these intimate connected gardens include several examples of edible plants. The Dry Garden has potted papayas, fig trees, and quince, among other warmth-loving plants. Here you will also see examples of taro, widely cultivated and native to Southeast Asia (though an important food staple in Africa and Oceania, too). You'll recognize them by their large, drooping leaves, reminiscent of elephant ears. The nearby Herb Garden, with its great variety of species (well over one hundred), contains plants that have been used for cooking, healing, ornamentation, or religious observance through the ages.

Though grand in terms of space and views, Wave Hill has intimacy and charm, unlike most institutional botanic gardens. And there are breathtaking views to enjoy wherever you walk. You will want to explore the grounds at your leisure, walking along meandering brick walkways, taking you from one spot to the next (you will find a convenient map of the grounds at the entrance), and enjoying the outdoor sculptures. On nice days you are likely to see people sitting in the grass or in comfortable wooden chairs scattered about the lawn, or photographing, sketching, or exploring.

Besides flowers and herbs and other edibles, you will find the Rock Garden (with tiny alpine plants and miniature flowers); the Wild Garden (so called because it has no hybrids); Aquatic Garden (water lilies and ornamental grasses); and the Monocot Garden (which displays a single group of plants in its variations). You might be surprised to find in this elegant setting a colorful splash of sunflowers and a pumpkin patch. But there they are, near the museum! Finally, be sure to stop in to visit the greenhouses and conservatory, a special treat in winter, when they provide a welcome touch of spring.

Wave Hill periodically hosts events, conferences, and workshops, sometimes related to food; a recent example was a "Honey Weekend," featuring honey-extraction demonstrations and cooking with honey (the estate has hives and "resident bees"). You'll find there are many things to see, do, and savor at this remarkable site.

18·
CHERRY TREES AND HERBS

Plantings from Around the World at the
Brooklyn Botanic Garden, Brooklyn

✍ DIRECTIONS

1000 Washington Avenue, Brooklyn, NY. Take the Manhattan Bridge, whose continuation in Brooklyn is Flatbush Avenue; stay on Flatbush all the way to Grand Army Plaza at Prospect Park and take the rotary three-fourths of the way around to Eastern Parkway, which borders the park. The Botanic Garden is immediately after the Central Library building. There is a large parking area (small fee). By subway: Take the B or Q to Prospect Park or the 2 or 3 to Eastern Parkway.

✍ INFORMATION

Open May–August, Tuesday–Friday, 8 A.M.–6 P.M., and weekends and holidays, 10 A.M.–6 P.M.; September–April, Tuesday–Friday, 8 A.M.–4:30 P.M., and weekends and holidays, 10 A.M.–4:30 P.M. Conservatory hours: Tuesday–Friday, 10 A.M.–4 P.M., and weekends and holidays, 11 A.M.–4 P.M. Telephone: 718-623-7200. Web site: www.bbg.org.

✍ The Brooklyn Botanic Garden is one of those surprises you happen upon in New York. In the midst of busy urban sprawl, around the corner from a dreary stretch of Flatbush Avenue (but near the lovely Prospect Park), you enter the iron gates of the Brooklyn Botanic Garden. There you find yourself in an enchanting, colorful, and completely intriguing world of planned gardens, elegant walkways, weeping cherry trees, and the many sights and smells of the world's most inviting gardens. The area was reclaimed from a waste dump in 1910. It takes up some fifty acres (but seems much larger), and you can walk around the property quite randomly, from the Japanese paths along a lake to the former rose gardens, from the Shakespeare Garden to the excellent conservatories. There are many pleasures in these fifty acres, particularly if you take this outing in spring.

Every season highlights a different area or style of garden, but surely April, May, and June are the most colorful times to visit, when the ornamental trees, luxuriant roses, and many spring flowers are in bloom. But the rock garden is ablaze with flowers during the entire growing season, and different species of roses bloom through September. A fragrance garden, labeled in Braille, is another fine section of the gardens; it too is open during the spring, summer, and fall.

All the plants are labeled, and there are more than twelve thousand of them. The conservatories and outdoor gardens include plants from almost every country in the world. If your taste is for literary references, you can enjoy the Shakespeare Garden, where plantings are related to passages from the Bard's works. If you want to meditate, you might choose to sit along the banks of the Japanese Garden's lovely walkways. If you are a horticulture fan, there is a section with local flora and many interesting displays of temperate, rain forest, and desert plants.

At the two main entrances to the Botanic Garden (on Washington Avenue), you can pick up a useful map, which will point you in the right direction. A good place to begin your walk is the Herb Garden, near the parking lot. This charming contoured plot contains more than three hundred carefully labeled herbs that have been used for medicine and cooking since the Middle Ages. Intricate Elizabethan knots form an intriguing pattern amid the plantings and add a unique element to this garden. In fact, the Herb Garden is now a centerpiece of the gardens, featuring herbs from around the world. Anyone with a particular interest in herbs should not miss this display. (There are also vegetables here.)

From the Herb Garden, you can take a lower or an upper walkway. The upper path will lead you to the Overlook, bordered by ginkgo trees, and to the grassy terrace known as the Osbourne Section, where a promenade of green lawns with stylishly shaped shrubs and freestanding columns await you. The pleasant, leafy lower lane will take you past grouping of peonies, crab apple trees, and wisterias to the Cherry Esplanade. We recommend that you see this garden in late April or early May, when the deep pink blossoms of the "Kanzan" cherry trees are a breathtaking dreamlike pastel. The trees are arranged in rows alongside tall, red-leaved varieties of Norway maples, whose dark leaves create a wonderful contrast in color. Some experts consider this cherry tree area to be the most beautiful garden in America.

19·
FROM NATIVE AMERICAN CROPS
TO EDIBLE FLOWERS
The New York Botanical Garden, Bronx

✿ DIRECTIONS

2900 Southern Boulevard, Bronx, NY. By car: From Manhattan take the Henry Hudson Parkway north to Mosholu Parkway (exit 42). At the end turn right onto Kazimiroff Boulevard. Follow signs for the car entrance to the garden. There is a parking fee. By subway: Take the B, D, or 4 train to the Bedford Park Boulevard station. From the station exit either take the Bx 26 bus east to the garden's Mosholu Gate entrance or walk eight blocks down the hill on Bedford Park Boulevard to the end (about twenty minutes); turn left onto Kazimiroff Boulevard and walk one block to Mosholu Gate entrance. By train: Take the Metro-North Harlem line to the Botanical Garden stop.

✿ INFORMATION

Open year-round, Tuesday–Sunday (and Monday on federal holidays), 10 A.M.–6 P.M. Telephone: 718-817-8700. Web site: www.nybg.org.

✿ A visit to the New York Botanical Garden in the Bronx, where you are transported to a special world, will lift your spirits at any time of the year. For here, in this wonderful and vast oasis of natural beauty, all sorts of plants and flowers—some of them edible—proliferate among woodlands, ponds, brooks, hills, and gardens.

The Ruth Rhea Howell Family Garden, one of the many (about fifty) display gardens found here, is of particular interest. This one-and-a-half-acre garden, filled to the brim with vegetables, fruits, herbs, and flowers, is a hands-on facility to teach children about plants and how to take care of them. Classes and workshops are offered to introduce kids to the excitement of planting, weeding, watering, and harvesting. They also get to taste, to pot a plant to take home, to dig for worms, and to search for other visiting garden animals in nearby meadows.

The variety of plantings in this space, which is divided into ninety small plots, is both colorful and aromatic, including tomatoes, lettuce, chives, cabbage, cilantro, peas, Swiss chard, marigolds, lemon balm (for tea), Johnny jump-ups (for tea and for flavoring ice cubes in summer), and kohlrabi, a deep-violet root vegetable that looks like a very large radish (with a milder taste). There are also themed edible plantings: a Pizza Garden (we'll leave you to figure out what that is!) and a Breakfast Bowl garden, including rye, corn, barley, and oats.

The Best Pumpkin Soup

Serves 8

2½ pounds pumpkin, peeled and cut into cubes
5 cups chicken stock
1 cup chopped onion
¾ cup white part of scallions
2 cups light cream
Salt and pepper to taste
8 thin slices of red-ripe tomatoes
1 cup unsweetened heavy cream, whipped
¾ cup finely chopped green part of scallions

In a large kettle combine the pumpkin, chicken stock, onion, and white part of scallions. Bring to a boil and simmer until the pumpkin is tender. Put the mixture through a fine sieve, or purée in an electric blender. Cool.

Stir the light cream into the soup, and season to taste with salt and pepper.

Pour the soup into eight chilled cups, and float a thin slice of tomato on each serving. Spoon the whipped cream onto each slice, or push it through a pastry tube. Garnish each serving with sprinklings of chopped scallions.

Another featured spot within the Howell Family Garden is the Lenape collection, presumably created on the site of a Native American Lenape settlement. (It was identified as such when buried relics from a nearby garbage dump were discovered.) The vegetables planted here are corn, squash, and beans, staples of the Lenape—and for that matter, modern American—diet. Children, from pre-kindergarten through elementary-school age, are welcome to enroll in weekly classes or occasional workshops offered here or to just drop by for a visit. Kids will also want to explore a three-foot-high hedge maze, as well as a recently added pumpkin patch to enjoy in fall, located directly across the street.

The New York Botanical Garden—one of the largest and most important botanic institutions in the country—was the creation of Nathaniel Lord Britton, a young American botanist. While on his honeymoon in England in 1889, he and his bride visited the Royal Botanic Gardens at Kew outside London. They were so inspired by what they saw that they were able to convince the Torrey Botanical Club in New York to create a similar public institution for botany and horticulture within the newly formed Bronx Park. The resulting complex is on a grand scale within some 250 acres of spectacular grounds.

You'll find that there is something for everyone here. For those who are especially interested in edible plants, there are other sites (besides the Howell Family Garden) to discover at the Botanical Garden. (Note: You can either walk from

New York Botanical Garden, Bronx

one site to the next using the handy garden map available at the entrance or take a tram, which passes every fifteen minutes or so.) One is the herbarium, the fourth largest in the world, featuring an archive of over seven million dried botanical specimens dating back more than three centuries; a section of it, the fungus herbarium, is the second largest in the Western Hemisphere, with about five hundred thousand specimens. Another is the remarkable collection of trees (some of them more than two hundred years old!) spread out over the fifty-acre native forest. Among them are quite a number of nut trees: pecans and hickories, including bitternut pignut, nutmeg, and shagbark; walnuts; hazelnuts, including American, European, Beaked, and Turkish; filberts; chestnuts, including Japanese, American, European, and Chinese; and gingko trees.

Finally, don't miss the incredible Conifer Arboretum, the Rock Garden, the brand new Azalea Garden, the Peggy Rockefeller Rose Garden, the Perennial Garden, the exquisite orchid collection, and, of course, the Enid A. Haupt Conservatory. Named a New York City Landmark in 1973 and recently restored, the conservatory is the nation's grandest and largest Victorian-style glasshouse. (The structure was ambitiously patterned after the Palmer House at Kew Gardens and the Crystal Palace at Hyde Park in London. It has been and continues to be one of the main attractions of the garden since its beginnings.) Here you will be fascinated by displays of tropical rain forests, carnivorous and aquatic plants, an enormous collection of palm trees, rainforest and desert plants, and beautiful seasonal flower shows. You'll find that there is something for everyone here.

The Latest on Greenhouses

Greenhouses come in a variety of styles, ranging from the simplest structure to the most elegant glass enclosure filled with ravishing exotic flowers; but their function remains the same: to protect whatever grows inside. To this end, new design technologies have been adopted, to work best in areas where the climate is more moderate. In eastern Long Island, instead of traditional greenhouses, you might see curious-looking structures called "hoop houses." These polyurethane-covered circular forms generally use double layers of plastic (with a few inches of "dead space" between, as insulation), to keep interior temperatures at the ideal level for year-round crops.

There are several reasons why plastic greenhouses are so popular. Farmers can better meet the growing demand for local produce throughout the year, which is obviously more ecological, fresher, and of higher quality; plastic structures are much cheaper than glass and are considered "temporary" in some areas and are thus tax exempt; they can be easily moved from one place to the next as necessary, to ensure the best protection of crops.

These days the consumer can savor just about any fruit or vegetable throughout the year, largely because of greenhouses. For example, juicy, flavorful "summer" tomatoes can now be found year-round in the equivalent hydroponic tomatoes, grown in water inside greenhouses (and really quite delicious). Meanwhile, new technologies are being considered, challenging the role of the greenhouse. Among them is one in which outdoor crops are covered with the double protection of a floating row cover, in addition to perforated plastic sheeting. This method has been highly successful in northern Europe, as it raises the temperature of the soil, enabling plants to grow faster. Who knows what will come next? Virtual greenhouses?

20·
HOW TO DO IT YOURSELF

Demo-Gardens, Programs, and New Ideas at the Queens Botanical Garden, Flushing

✒ DIRECTIONS

43-50 Main Street, Flushing, NY. Take the Long Island Expressway east to exit 23 (Main Street); go north to the corner of Main and Dahlia. The entrance is on the left. By train: Take the no. 7 subway or Long Island Rail Road to Main Street/Flushing; then take Q44 or Q20 bus or walk eight blocks south to the Queens Botanical Garden.

✒ INFORMATION

Open daily, but call ahead to find out when demonstrations are scheduled. Telephone: 718-886-3800. Web site: www.queensbotanical.org.

✒ The Queens Botanical Garden is near the 1939 World's Fair site; it was once a garbage dump. But since 1963, its thirty-eight acres have been elegantly designed into a series of individual gardens, including a rose garden, a bee garden, a rock garden, a children's garden, a fragrance walk, and so on. There are some four hundred types of petunias, a garden of heather, an area of azaleas, walks through woodlands and wetlands, and many more pleasures to enjoy right in the middle of an urban environment.

Of particular interest to us are the herb garden, the bee garden, the vegetable garden, and the demonstrations of how the knowledgeable gardeners go about creating and tending them. This educationally oriented botanical garden seems to specialize in many subjects useful to both the gardener and the

---

Jeannie's Spring Soup Recipe

Slice some mushrooms and some scallions, and sauté them in butter until nearly done. Add some fresh chopped spinach, and sauté a little longer. Cook some fresh chopped asparagus in chicken stock until nearly done. Add asparagus to sauté mixture, salt and pepper to taste, and cook another minute or so. (Optional: add a little cream after cooking has stopped.) Serve immediately, sprinkling a few freshly chopped scallions and some nutmeg in each bowl. Note: Amounts aren't important. Put in more of what you like and less of what you don't.

environmentalist; there are demonstration gardens of bioswales (depressions in the ground that collect rainwater that runs off into the nearby areas), compost demonstration gardens, an exhibit of the recycling of gray water into "constructed wetlands," and even green roof planting.

Among the many specific gardens is one called Backyard Gardens, where "small-scale" gardening techniques are displayed. (These smallish areas were originally designed and maintained by various nurseries in the area.) In the Bee Garden you can learn what plants attract bees or flavor honey; beehives are nearby. In the Herb Garden there is a collection of woody, annual, and perennial plants that are identified as to their uses: culinary, medicinal, dyeing, aromatic.

In the staff vegetable plots, vegetables and berry bushes are intertwined with topiary from a former time. On certain weekends during growing season, experts demonstrate their techniques, and anyone who enjoys gardening will find these programs filled with tips and new ideas.

21·
A BUCOLIC OUTPOST IN
THE MIDDLE OF QUEENS

*The Surprising Queens County Farm Museum,
Floral Park*

✿ DIRECTIONS

73-50 Little Neck Parkway, Floral Park, NY. Take the Triborough Bridge to Grand Central Parkway east. Take exit 24 south (Little Neck Parkway) for three blocks. The museum entrance is on your right. (The entire trip is about a half hour from Manhattan.)

✿ INFORMATION

Open daily, 10 A.M.–5 P.M.; guided tours Sundays; many special events. Telephone: 718-347-3276. Web site: www.queensfarm.org.

✿ You may think of Queens as part of one of the great urban centers of the world. But here just a few minutes from the highways is a large (forty-seven acres) working farm that is lovely to meander through, has a terrific history, and actually produces fruits and vegetables. We could hardly believe our good

Queens County Farm Museum, Floral Park

fortune on finding such a spot—and then we discovered that numerous schools know all about it, and children on hayrides were all around us!

First a word about its history. This farm is the city's largest remaining tract of farmland and is a significant vestige of three centuries of agriculture, In fact, it was founded in 1697. The land was sold to a Dutch farmer, Elbert Adriance, and it remained a Dutch-owned farm for the next two hundred years. You can see the farmhouse (well kept on the premises, with its original characteristics and additions over the years). It has been restored to its appearance in 1855. A series of families ran the farm until 1975, when it became the Queens County Farm Museum.

But the agricultural work has continued, uninterrupted, with the farm producing everything from grapes for wine to vegetables, corn (there is a large cornfield and a corn maze in fall, the only one in New York City), many farm animals, and fields of pumpkins, herbs, tomatoes, apples, and so on. Unlike most working farms, this is a place that wants visitors to enjoy it, so you can walk all around, visit the historic house, and watch a variety of farm chores being performed.

Ideally suited for children in tow, the farm runs every sort of event, from candle making with its own honey wax to painting pumpkins, harvest celebrations, and a variety of fairs. (We should add that a part of the farm is devoted to children; called Green Meadows Farm, it is in the southern part of the museum and is host to these many events, while the historic portion with its antique farmhouse is a quieter growing section.)

22 ·
ONE OF NEW YORK CITY'S LAST WORKING FARMS

Decker Farm in Historic Richmond Town, Staten Island

⌘ DIRECTIONS

435 Richmond Hill Road, Staten Island, NY (New Springville section of Staten Island). Take the Verrazano Narrows Bridge to Staten Island. Follow I-278 west to Richmond Road/Clove Road exit. At the second light, turn left on Richmond Road. Go four and a half miles to end and make a left onto Arthur Kill Road. Go one block to Clarke Avenue and turn left. Parking is on the left.

⌘ INFORMATION

Open Wednesday–Sunday; call for seasonal information and special events. Telephone: 718-351-1617. Web site: www.historicrichmondtown.org.

⌘ It's hard to imagine New York City having working farms, but once there were many of them, particularly in the outer boroughs. A recent conservation effort to save such places succeeded on Staten Island, where Decker Farm, a part of Historic Richmond Town (an antique village with old-time activities) has been protected from development by the Trust for Public Land and the Staten Island Historical Society, among other organizations.

The farm has a long history, beginning in 1809, when it began operation; the farmhouse is still standing. The Decker family owned it from 1841 until 1955, when it was bequeathed to the Staten Island Historical Society.

Along with farm activities long performed at the site (beating rugs, raking hay, carrying water with buckets and yokes, for example), you can see rows of vegetables (four kinds of lettuce, butternut squash, pumpkins, zucchini, and corn, among other crops). The farm uses organic methods. This is an educational facility, and visitors can enjoy a variety of public programs, including taking part in the old-fashioned chores. (How about washing clothes on an eighteenth-century washboard?) In addition, a farmers' development program exists here, and students of farming are instructed in organic methods.

New York's Premier Farmers' Market:
Union Square Greenmarket

Looking to buy the tiniest microveggies, the freshest lemongrass, or the greatest variety of edible flowers? Union Square Greenmarket is your place! Not only is this the largest market in the region; it is also the most unusual and innovative. Operating on about two acres of parkland at Midtown Manhattan's Union Square, it hosts at least 150 weekly vendors who offer over one thousand varieties of fruits and vegetables. With as many as 250,000 customers per week the market has also become an important tourist attraction for the city—and a great spot for people watching! You can find other items here besides produce, such as hard-to-find meats and fish, artisan breads, farmstead cheeses, jams, pickles, cut flowers and plants, wine, ciders, and maple syrup. If that weren't enough, special events are held on this bustling site: cooking demonstrations, book signings, and a variety of how-tos relating to farming and preparing food. There is something for everyone here, from top chefs to local kids wanting to grab an after-school treat. (Yes, there are also doughnuts here!) Started in 1976 with just a handful of small family farmers, this open-air wonder has grown prodigiously. Another notable thing about this well-run operation: many of the farmers donate end-of-day unsold produce to local food pantries and soup kitchens.

Open year-round, Monday, Wednesday, Friday, and Saturday, 8 A.M.–6 P.M. Union Square Park, Fifteenth–Seventeenth Streets and Broadway, New York, NY. Telephone: 212-788-7476.

NEW YORK STATE
Long Island

23 ·

UNEXPECTED DELIGHTS

Tropical Delicacies in the Greenhouses of Planting Fields Arboretum, Oyster Bay

❧ DIRECTIONS

1395 Planting Fields Road, Oyster Bay, NY. Take the Long Island Expressway east to exit 39 north (Guinea Woods Road), to Route 25A (also called Northern Boulevard). Turn right on Route 25A, pass C. W. Post College, and make a left on Wolver Hollow Road (Route 4). Planting Fields Road will be a right turn after about two and a half miles, and the entrance is on the right.

❧ INFORMATION

Open November–mid-April, daily, 10 A.M.–4 P.M. There is an admission fee. Telephone: 516-922-9200. Web site: www.plantingfields.org.

❧ An arboretum is not just a collection of trees. Planting Fields is a horticultural center for people who enjoy every sort of growing things; there are greenhouses, plant collections, a wildflower walk, trees, nature trails, sweeping lawns, a dwarf conifer garden, and in springtime a marvelous profusion of azaleas. In various seasons the 409-acre estate offers just about every sort of plantlife environment: about 160 acres are developed woods.

This was once the private estate of William Robertson Coe. His home, Coe Hall, an elegant example of Elizabethan-style architecture, is occasionally used for concerts, other cultural events, and school activities. The grounds are meticulously kept (in the developed section), with changing displays and identifying tags on trees, shrubbery, and flower beds. The wilder part is crisscrossed with trails. The arboretum is divided into several sections. You can plan to do them all or focus on separate parts. Here is what you can see outdoors:

The grounds have majestic trees, many of which are the largest of their kind on Long Island. There are azaleas and rhododendrons galore. Some six hundred plants make up this collection, which is famous throughout the East.

Of particular interest to us in our search for nature's bounty is an extraordinary greenhouse, known as Hibiscus House. Here, in addition to brilliant displays of flowers—familiar and exotic—is a most unusual collection of food plants. You'll see tropical specialties, including coffee and cocoa plants, bananas, lemons, avocados—all thriving. We recommend this visit to families with children, who will be surprised to see how such basic foods—bananas and coffee, for example—look in their (almost) natural habitats. This is one of the only venues in the area where you can see foods native to warmer climates—don't miss it!

24 ·
SAVORING A DAY IN WINE COUNTRY

Seven of the North Fork's Best Vineyards, Cutchogue, Peconic Region

ᔎ DIRECTIONS

Paumanok Vineyards: *1074 Main Road, Route 25, Aquebogue, NY. Take the Long Island Expressway east to exit 73 and take Route 58, which becomes Route 25 east (also called Main Road).*

Pellegrini Vineyards: *23005 Main Road, Route 25, Cutchogue, NY. Pellegrini Vineyards is in the next town, Cutchogue.*

Peconic Bay Winery: *31320 Main Road, Route 25, Cutchogue, NY.*

Pugliese Vineyards: *34515 Main Road, Route 25, Cutchogue, NY.*

Bedell Cellars: *36225 Main Road, Route 25, Cutchogue, NY.*

Pindar Vineyards: *37645 Main Road, Route 25, Peconic, NY.*

Lenz Winery: *38355 Main Road, Route 25, Peconic, NY.*

Shinn Estate Vineyards: *2000 Oregon Road, Mattituck, NY.*

ᔎ INFORMATION

There are several tours offered to North Fork's vineyards, including LongIslandTours .com and LongIslandWineTours.com. There are also many events scheduled at the vineyards, including tastings and winemaking. You can take a trolley ride (NorthFork Trolley.com) from vineyard to vineyard. Most vineyards invite you to sample their wines, and you will find a list of events and festivals for the public on their Web sites. We recommend visiting in late spring through early fall.

Paumanok Vineyards: Open summer, daily, 11 A.M.–6 P.M.,; winter, daily, 11 A.M.–5 P.M. Telephone: 631-722-8800. Web site: www.paumanok.com.

Pellegrini Vineyards: Open daily, 11 A.M.–5 P.M. Telephone: 631-734-4111. Web site: www.pellegrinivineyards.com.

Peconic Bay Winery: Open summer, Sunday–Thursday, 11 A.M.–5 P.M. Telephone: 631-734-7361. Web site: www.peconicbaywinery.com.

Pugliese Vineyards: Open daily, 11 A.M.–5 P.M. Telephone: 631-734-4057. Web site: www .pugliesevineyards.com.

Bedell Cellars: Open daily, 11 A.M.–5 P.M.; summer, Friday–Sunday, 11 A.M.–6 P.M. Telephone: 631-734-7537. Web site: www.bedellcellars.com.

Pindar Vineyards: Open daily, 11 A.M.–6 P.M. Telephone: 631-734-6200. Web site: www .pindar.net.

Lenz Winery: Open June–October, daily, 10 A.M.–6 P.M.; November–May, daily, 10 A.M.–5 P.M. Telephone: 631-734-6010. Web site: www.lenzwine.com.

Shinn Estate Vineyards: Open daily, noon–5 P.M. Telephone: 631-804-0367. Web site: www.shinnestatevineyards.com.

There are some forty vineyards on Long Island, a burgeoning and delightful industry in this rural area. A visit to a vineyard on a bright summer or early fall day is a treat—among our favorite outings.

Strolling through a vineyard, particularly at harvest time, can be very appealing. Because grapes need as much sun as possible, vineyards tend to be located in valleys of hillsides with the maximum sun exposure. They are usually open, airy places in a smiling landscape, and a walk through them can be most pleasant. It is particularly satisfying to go in late summer or early fall, when the mature plants are laden with plump fruit ready to be picked and smelling delicious. It is a time of abundance, when the visitor is especially aware of nature's growing cycle. The grapes will soon be picked and turned into wine. The atmosphere brings to mind ancient rites of harvesting and winemaking.

Although it is certainly informative to be guided through a vineyard on a tour and to learn about the elaborate process of winemaking, we enjoyed wandering about at will and getting the feel of what it's like. Some vineyard owners allow this, while others are reluctant to do so. With the burgeoning interest in wines, this region—particularly Long Island—has become dotted with vineyards large and small.

The North Fork is about 120 miles from New York City and, until it became a center for vineyards, was more a farming area than a mecca for tourists like the South Fork and the Hamptons. But since the development of the wine industry, many changes have taken place here, and you'll find restaurants, lodging, and a variety of events in the small towns that dot the North Fork.

Not since the Napa Valley vineyards surprised the wine world has a new area of vineyards been so celebrated. In the past thirty years these vineyards have begun producing sparkling wines, white wines, reds, and dessert wines (with particular emphasis on prize-winning chardonnays, merlots, and Rieslings).

We recommend visiting our favorites (in the list that follows) and walking through the orderly rows of growing vines, as well as enjoying each vineyard's different ambiance. Some offer outdoor tables for tasting, some take you to see how the wine is made, and others delight with the scenic beauty of their hillsides. In each you'll find a rare way to experience nature's bounty.

Paumanok Vineyards

Among our favorites, Paumanok Vineyards sits on seventy-seven scenic acres of rolling landscape, with beautiful views all around. As you drive in, you first notice a renovated turn-of-the-century barn, with tasting room, winery, and cellar.

From its attractive deck you can see flowering shrubs and the lush green vines stretching before you. The mood is inviting and welcoming.

Paumanok is a family enterprise. The Massouds—Charles, son of a Lebanese family of restaurateurs, and Ursula, daughter of a German family of vintners and brewers—were both raised with a deep appreciation for the land and its bounty. They started their own vineyard in 1983. The vineyard is entirely owned and managed by them, with help from their three sons, who are apparently as committed to vineyard life as they are. All are involved in every aspect of the business, from the harvesting of grapes to sales and marketing. As one of the sons puts it, all the work involved "is worth it when you have a great vintage and you have great customers who appreciate what you do."

Paumanok's list of award-winning wines—from the tasty cabernet sauvignons (among the top rated of this popular varietal in the state) and merlots to excellent sauvignon blancs, Rieslings, and chardonnays—is impressive. A premium red blend, Assemblage, has also received much praise. You will no doubt enjoy a sample or two in the tasting room before or after you take your walk in the vineyard itself. The Massouds are informal and allow visitors to walk on their own among the long rows of vines, which seem to stretch forever, savoring the scent of the grapes and abundant greenery.

Pellegrini Vineyards

The Pellegrini Vineyards are in Cutchogue, which claims to be the sunniest town in New York State—thus an ideal place for making wine. (Apparently its climate is similar to that of Bordeaux in France.) In fact, there are quite a number of premier vineyards in this small area, lined up next to one another off the main road. Pellegrini sits on a slight hill; its tasteful architecture blends well with the surrounding landscape, creating an aura of modern elegance, with charm. The main building is graced by a white-columned courtyard (where people sometimes sit to enjoy a taste of wine); if you look toward the abundant grapevines, you'll see a delightful white gazebo, completely surrounded by lush greenery. (Not surprisingly, the gazebo is sometimes used for weddings.)

Inside the main building is a spacious tasting room with a high, vaulted ceiling; go up a flight of stairs to the mezzanine for a wonderful view of the vineyards. (For a close-up view—before your outdoor walk—find the open deck.)

The vineyard was established in 1991 by the Pellegrini family, who came to the area in the early 1980s. Though winemaking had not been part of their family tradition, Bob and Joyce Pellegrini had a genuine interest in its many aspects —enough to start a collection of fine wines, to join wine clubs, and to read everything they could about wine before they actually plunged into making it. In the early days, they worked in the vineyards themselves. A year later they hired Russell Hearn, an Australian winemaker, to take charge of the winery. From

1992 (the vineyard's first vintage), Hearn has been actively involved as wine-maker and has been responsible for the types of wines produced.

Among the wines to look for are the succulent cabernet sauvignon (especially the 2004 vintage), the full-bodied 2004 merlot, two very tasty red blends called Reserve and Encore, and a host of other varieties. As you sample the wines, you can ask for recommendations, according to your taste. (The Pellegrinis or one of their staff is always available.)

A walk through the vineyard itself is not to be missed. The grounds are vast (about seventy acres are planted) and are a pleasure to explore and experience. You can come and go at will, or ask to be guided through.

Peconic Bay Winery

Peconic Bay Winery sits on a large (two hundred planted acres) and quite flat property, across the road from a shopping center. Though not as picturesque as some of the other vineyards we visited, it nonetheless plays an important role in the winemaking of the region. The production here (six thousand to eight thousand cases per year) is relatively small, making hand harvesting feasible. In fact, the staff is committed to this time-consuming and expensive practice, which results in the careful selection of only the best grapes during harvesting.

This is also among the oldest vineyards in the area. The first vines were planted in 1979 on what had been an early twentieth-century farm with a Dutch-style barn, now renovated and used as the tasting house. The present wine-maker, Greg Gove, said that winemaking is somewhat like raising a child: "You start with the grapes, and you watch the evolution of that wine all the way to the bottle, and you see how it changes. And when it does everything you wanted it to do, you feel very proud that the wine has evolved in the way you assumed it would."

Peconic Bay has won many awards for its wines, including for the 2005 cabernet franc, the 2002 cabernet sauvignon, the Riesling (voted one of the best in the country), the merlot (voted as the best in the state), and more. Enjoy a sampling in the appealingly light-filled tasting room with its attractive and long bar.

Tours of Long Island's Wine Country

All kinds of tours of this burgeoning vineyard region are offered. Try Long Island Vineyard Tours, which takes you on tastings (in a limo, no less) (telephone: 718-WINE-TOURS) or the North Fork Trolley Company, a picturesque old-fashioned red trolley that goes to both North and South Fork vineyards (telephone: 631-369-3031).

A walk through the vineyard can stretch for quite a distance (the flat terrain makes it seem even longer!), but the going is easy. Everyone is very friendly here, so feel free to ask questions if you're interested in a particular aspect of winemaking.

Pugliese Vineyards

Pugliese is perhaps the most idyllic-looking vineyard we have discovered in our explorations. What first catches your eye as you pull into the driveway is an unusually long arbor covered with luscious-looking grapevines; garden chairs are placed beneath it for people to sit in the cool shade and enjoy their surroundings —and perhaps to taste some wine, too. To the right is a delightful pond with a decorative fountain in the middle, adding charm to the setting—and glamour to the inevitable wedding parties!

This vineyard, established in 1980 and headed by Ralph and Pat Pugliese, is also a family owned and operated business. One son makes the wine, another manages the vineyard, and yet another makes himself available when needed. A daughter, daughter-in-law, or nephew can be found behind the bar in the tasting room. On occasion, when things get especially busy during summer and early fall, even friends are there to help out.

The story began when the Puglieses decided to buy a small summer cottage on two acres, mostly to get away from city life; the property expanded substantially when they bought the farm next door. Soon they had planted a few acres of grapevines—but only for their own private use. But the operation became more successful than they expected, with more grapes than they could use, and one thing led to another. Today the vineyard encompasses about fifty acres, including several buildings and an impressive champagne cellar.

Over the years the wines produced here have received many prestigious awards, including one for the excellent champagne. There are now some twenty wines to choose from: ports, pinot grigios, chardonnays, cabernet sauvignons, Gewürztraminers, and many more. Some of the bottles are hand-painted by Pat herself, adding a personal touch to the operation. She has also designed a special label for other wines.

The vineyard itself is most inviting, and you won't want to miss walking through. The place is informal, and you are most welcome to explore at your own pace and leisure.

Bedell Cellars

Among the North Fork's best-known wineries and vineyards is Bedell Cellars. It also has one of the prettiest entrances we found, graced by a profusion of lavender and other well-tended plantings. The grounds—some fifty-seven acres are

Bedell Cellars, Cutchogue

planted—are lushly landscaped, inviting the visitor to explore. The buildings appear as traditional barns on the outside, with a chic and contemporary (and recently redone) interior of black, white, and chrome. A surprising contemporary art collection is featured on some walls. (The present owner, Michael Lynne, is apparently an avid art collector.) A spacious open deck overlooks acres of vines in tidy rows, as far as the eye can see.

Bedell dates back to the late seventies, when founder Kip Bedell, home winemaker on the side, discovered the North Fork and bought a few acres on what had been a potato farm. He cleared the land, planted vines, and by 1983 had transformed the property into a small vineyard. In 1986 he and his wife sold their first vintage, using a rustic picnic table in the old potato barn. Those were the early days of winemaking in Long Island, when the few vintners there were still feeling their way around. Over the years Bedell and the others learned new techniques to fight humidity, to get more sunlight on the grapes, and to harvest more mature fruit. Much has changed since then—from expansion and renovation to installation of new equipment from Europe.

Bedell's wines have a well-deserved reputation. Their latest crowd pleaser is called Musee, which has received the highest accolades and scores. Chuck Close, the contemporary artist, designed its label. Another wine to look for is a white blend named Gallery. (Its label has also been designed by another contemporary artist, Ross Bleckner.) There are also many varietals, from chardonnay, Riesling, and Gewürztraminer to cabernet and merlot.

Pindar Vineyards

Pindar is not only Long Island's largest vineyard (with almost seven hundred acres of vines spread out throughout the area) but also its oldest under continuous family ownership. Its vast grounds include a large tasting pavilion adorned with white columns, a red-tiled roof, and a Tiffany-style stained-glass window; a spacious pavilion in the back for sampling wine; an exceptional winery with all the latest accoutrements (temperature-controlled steel tanks, etc.); and a climate-controlled cellar holding at least three thousand barrels, with an elaborate machine that turns hundreds of bottles of wine at a time.

Pindar was established in 1979 on thirty acres of farmland by Herodotus Damianos, a Long Island physician, and his family. Damianos, a wine connoisseur from Greece, had always been interested in wine and agriculture, and he joined the early vintners on the North Fork. Over the years the vineyard thrived, with the help of his three sons.

The list of wines made here is long and varied, from the simplest to the most complex and subtle. The family likes to experiment and plant different varieties and try new things. The result is that there is quite a bit of innovation. Recently they've come up with late-harvest varieties—from sauvignon blanc to chardonnay. But the best known of Pindar's wines is Mythology, a very popular wine, considered by some critics to be among the top fifty in the world.

The environment at Pindar is open and friendly, and you'll enjoy visiting here. There are long walks to take around the spacious grounds, and you are welcome to explore at will, though complimentary guided tours are also offered.

Lenz Winery

Lenz Winery is somewhat reminiscent of a French country property in the Bordeaux region, with its mansard roof, charming flower-filled arbor, outdoor decorative wine barrels topped with potted plants, picnic tables, pretty courtyard, restored farmhouse, and acres of luscious greenery. The inviting look of the place is picturesque and rustic, while its winemaking operation is, in fact, state of the art, with the latest technology in place.

Lenz was established in 1978; its seventy or so planted acres contain some of the most mature vines on Long Island. Peter Carroll (a former banker) and his wife, Deborah, were both wine enthusiasts when they bought the winery a few years later, but they were hardly experts. Like many other vintners in the region, they learned as they went along and engaged a fine winemaker, Eric Fry, who had worked in California and in France. Together, their efforts have paid off, judging by the quality of wines they have produced and the accolades they have received.

The wines here have done consistently well. The reds have been compared to the best of France (in a 2006 controlled blind tasting, even besting the great

Lenz Winery, Peconic

French Chateau Figeac). The white and sparkling wines have also had impressive results. You can walk through the vineyard and watch the various aspects of winemaking (see sidebar "Do-It-Yourself Winemaking").

Shinn Estate Vineyards

The most forward-looking aspect of Long Island's North Fork vineyards can be seen at Shinn Estate Vineyards. With a working philosophy of sustainability and low-impact farming, the owners are pioneering the use of wind and sun for producing all the electricity they need for their winery operations. They have a wind turbine and several solar panels, and innovative soil building. You can spend a night at their recently restored 1880s farmhouse, which serves as a bed-and-breakfast.

Do-It-Yourself Winemaking

Once a year you can try your hand at winemaking at Lenz Winery on Long Island's North Fork. Thirty individuals (call for a reservation) join on two different days (fifteen each day) to experience barrel tasting and wine blending, among other aspects of the craft.

In connection with the process of winemaking itself, Lenz offers something special to its visitors: a chance at "making wine," or sampling and judging the wine before it is actually completed. On given afternoons during the year (call for information) guests are invited to stand in a dimly lit, basement-like room with a cement floor, surrounded by enormous stainless-steel tanks filled with wine. The wine is poured in glasses and carefully considered. This demonstration is a great learning experience for anyone interested in the art (and science) of winemaking. Telephone: 631-734-6010. Web site: www.lenzwine.com.

A small vineyard in Peconic, Sannino's Bella Vita Vineyard, offers another opportunity. If you'd like to try your hand at making wine from the vineyard's grapes, you can sign up for "Vine to Wine," a course that starts with choosing the grapes and goes all the way to the finished product. Sannino's Bella Vita Vineyard is located at 1375 Peconic Lane in Peconic (Long Island). Telephone: 631-734-8282. Web site: www.sanninovineyard.com.

Additional places that offer comparable adventures in do-it-yourself winemaking include Staten Island Winery, telephone: 718-494-9463, Web site: www.siwinery.com; Brooklyn Winery, telephone: 347-763-1506, Web site, www.bkwinery.com; Make Your Own Wine, Westchester's School of Winemaking, Elmsford, NY, telephone: 914-741-5425, Web site: www.myowine.com.

25 ·

A BIODYNAMIC FARMER COMBINES NATURE AND COSMIC FORCES

An Unusual Experience at "The Farm," Southold

✍ DIRECTIONS

59945 Main Road, Route 25, Southold, NY. From the Long Island Expressway at Riverhead, take Route 43 to Route 25 east to Southold. The Farm is on Route 25, on your right in Southold.

✍ INFORMATION

Open daily, 8 A.M.–5 P.M. Visit any time in growing season. Look for vegetables, flowers, and beehives. Telephone: 631-765-2075.

✍ This is one of the more unusual farm experiences we've had. We have visited many examples of nature's bounty but none in which the proprietor has such a personal relationship with her plants!

In fact, K. K. Haspel, the owner, speaks to her plants on a regular basis and consults with them about what to plant, where, and when. Using a divining rod (a forked branch or stick that indicates subterranean water or minerals by bending downward when held over a source), she determines where to place her vegetables—often contiguously with flowers, which she claims is mutually beneficial. (As an example, she showed us tomatoes growing next to lettuce and beans and rows of poppies and zinnias.)

Haspel also believes in the benefit of insects and rejects the use of pesticides. "Weak plants send off a smell," she says, "while healthy ones don't send that signal, and the pests don't come. When you have a balanced ecosystem, . . . everything is healthy." As you walk around with her, you will quickly find yourself yielding to the otherworldly aura of The Farm.

The story of The Farm begins in 1999, when Haspel and her husband, Ira, were searching for a country property. They found these five acres, which had been abandoned for years, and decided to buy the property. K. K. Haspel claims to have had a vision in which she saw an organic farm at the site, based on the biodynamic farming principles of Rudolph Steiner.

Rudolph Steiner (1861–1925) was a German philosopher and occultist who founded a movement he called Anthroposophy, in which he emphasized that nature responds to the light and heat of otherworldly forces and that all the planets are interrelated. He pioneered biodynamic growing techniques, including mixing compatible crops and regular composting. (Farming techniques based on Steiner's views are said to produce four times the output of ordinary farms.) In addition to Steiner's biodynamic teachings, Haspel uses the Stella Natura calendar, which indicates special days for planting, according to the alignment of the stars and planets.

When you visit The Farm, you'll be dazzled by the colors of flowers interspersed throughout the long raised rows of very robust-looking vegetables. There are some twenty different crops, including potatoes, dozens of varieties of tomatoes, spinach, kale, leeks, beets, garlic, and much more, as well as the flowers: zinnias, sunflowers, poppies, and so on. Everything is grown from organic heritage seeds. If you are a gardener yourself, you will enjoy visiting the working greenhouse with its organic composting system. In addition, ask to see the thriving beehive, which produces eighty to one hundred pounds of honey a year; the bees have a three-mile radius to The Farm's flowers.

Despite its uncommercial attitude, this is a very successful enterprise. Many tourists buy Haspel's produce, as do local restaurants. With her enthusiasm and commitment to biodynamic techniques, this unofficial ambassador of sustainable farming makes growing things seem a pleasure rather than drudgery. You will leave feeling upbeat, your arms filled with flowers and vegetables that you know were grown with a special relationship to the farmer.

26·
A FEAST FOR THE SENSES

Lavender by the Bay, East Marion

🌺 DIRECTIONS

7540 Main Street, East Marion, NY. Take the Long Island Expressway east all the way to exit 73 (Route 58), onto Old Country Road, toward Orient/Greenport; after four miles take Route 25 east for about half a mile; turn right on Sound Avenue and travel for five and a third miles; continue on Middle Road for nine and a half miles; continue to follow Route 48 east for three and three-quarters miles; continue on Route 25 for a mile and a half. The site will be on your right.

🌺 INFORMATION

Open May 1–October 31, 9 A.M.–5 P.M.; during the summer season, open daily; for most of September and all of October, open only Friday–Sunday (9 A.M.–5 P.M.). June is especially recommended. You can pick your own lavender from June through September. Telephone: 631-477-1019 or 917-251-4642. Web site: www.lavenderbythebay.com.

🌺 If you've never been through a field of lavender on a summer day, here is your opportunity for a blissful experience. When the lavender is in full bloom —its purple, blue, white, or pink flowers swaying in the gentle breeze—the air is intoxicating with delicious fragrance. The scene is reminiscent of an impressionist painting.

Lavender by the Bay, East Marion

Lavender by the Bay, on the east end of Long Island's North Fork, features more than twenty thousand lavender shrubs of many varieties, planted on seven acres. It is a peaceful site, blessed with just the right exposure to the sun and relatively moderate temperatures—ideal for growing this lovely plant. You are welcome to walk (carefully) among the abundant rows, to photograph, to pick your own flowers (June through September), or simply to look and enjoy the view. If you come before the full blooming season, you will see long stems with thin leaves—hardly worth the effort! But come May—and well into September —you will be amply rewarded: the evergreen shrubs grow to three or four feet in height, and the flowers on them are bountiful.

Lavender is a remarkably versatile plant with a surprising number of uses. There are about twenty known species. Originating in the Mediterranean region, where it grew wild, it was prized by the ancient Greeks for its medicinal benefits. The Egyptians used it for mummification, and the Greeks and Romans bathed in lavender-scented water. Thanks to the Romans, lavender spread to the rest of Europe and became popular for alleviating rheumatism, rashes, infections, and colds. Queen Elizabeth I drank lavender tea to ease her terrible migraines. As part of the same family of herbs as thyme, basil, rosemary, and sage, it was (and continues to be) used to flavor soups, stews, and drinks. And then there is that powerful aroma, which still to this day enhances soaps, creams, perfumes, and cosmetics throughout much of the world. Lavender is also used in decoration. In contemporary high-end cuisine you sometimes see it on top of an elegant salad or cake. Be sure not to leave it untouched on your plate: it can be eaten and is quite tasty, too!

In fact, eating flowers is not a new thing at all: people have done it for centuries. In some of today's upscale restaurants, it has become very popular. These so-called edible flowers include not only lavender but also apple blossoms, dandelions, carnations, honeysuckle, impatiens, nasturtium, pansies, chives, and many other familiar names—the list is longer than you might expect. (Did you know that broccoli and cauliflower are flowers?)

After your walk, be sure to stop at the farm's attractive shop, where you will find an array of lavender goods: sachets, fresh cut or dried lavender, and lavender plants; it also sells honey from the farm's own honeybee hives (no doubt lavender honey!), which will remind you of your day at Lavender by the Bay.

27 ·
DESTINATION GARLIC
Visiting Biophilia Organic Farm, Jamesport

✿ DIRECTIONS

211 Manor Lane, Jamesport, NY. Take the Long Island Expressway east to exit 72, where you will pick up Route 25 east. Manor Lane intersects Route 25 in the center of Jamesport; go left (north) until you come to the farm.

✿ INFORMATION

Open June–October, Saturday–Sunday, 10 A.M.–6 P.M. Telephone: 631-722-2299.

✿ It may surprise you to know that there are some forty varieties of garlic. And many of them are grown right here on an organic farm and harvested around the Fourth of July. We loved the setting of this farm, with its fourteen acres of walking paths meandering through beds of garlic—as well as other crops (including herbs, figs, nine types of tomatoes, and flowers).

It is an unusual setting, bucolic and unspoiled—a terrific farm walk through fields and meadows. But it also appears to be thriving as a working farm. (People can buy shares here for growing their own crops; it accommodates thirty or forty shares.)

Garlic is, of course, related to the onion, with a strong odor and a bulb with a highly distinctive flavor. You'll find it interesting to see it growing in neat rows, with only slightly noticeable differences of color and size depending on the variety.

You can bring the kids on this outing, because the owners are very welcoming, the paths are lovely and clearly separate from the growing beds, and the kids will think farm life is nothing less than idyllic!

How to Plant Garlic

On a bright, sunny fall day not too long ago we watched a group of people (including two small children) planting garlic on a farm in Rockland County, New York. It looked like fun, and quite simple, too, something everyone in the family could enjoy doing together. Here is how you do it:

1. Garlic should be planted in the fall, a few weeks before the soil usually freezes. Choose a sunny spot on well-drained soil. (A raised bed is ideal.)

2. Plant the individual cloves of the garlic bulb. Each clove will grow into a bulb.

3. Using a hoe, make a two-inch-deep furrow in the soil. Place the cloves inside, about five inches apart in rows (twelve to eighteen inches long), with the pointy end of the clove end up. Bury the cloves with a six-inch layer of organic mulch.

4. Because garlic needs about an inch of water per week, you might need to water the area in case it doesn't rain. Look for the earliest green shoots to appear in four to six weeks. (They will stop growing once it gets cold.)

5. In spring, when you see more active growth, pull back the mulch and add one inch of compost. Replace the mulch around the garlic to retain moisture and to prevent weeds from growing.

6. The garlic can be harvested in the early summer, as the bottom leaves of the plants begin to turn yellow.

28·

PICKING EGGPLANTS AND PEPPERS AND FINDING YOUR WAY THROUGH A CORN MAZE

A Visit to F&W Schmitt's Farm, Melville

⊱ DIRECTIONS

26 Pinelawn Road, Melville, NY. Take the Long Island Expressway east to exit 49, Route 110. Go to the first light and turn right on Pinelawn Road. You can't miss the farm. Parking is on your right.

⊱ INFORMATION

Open Monday–Friday, 9 A.M.–6 P.M.; Saturday–Sunday, 9 A.M.–5 P.M. (During the fall season the farm is also open on certain nights between 7 P.M. and 11 P.M.) Telephone: 631-271-3276. Web site: www.schmittfarms.com.

⊱ This large and lively farm offers much for families with kids, especially in the fall season, when a lot goes on. Offerings include a wonderful pumpkin patch (pick your own pumpkins or just buy them), free hayrides, pony rides, farm animals, an enormous hay pyramid (we are told Long Island's largest), and a five-acre corn maze, which is sometimes "haunted" and changes theme each year. In summer and fall there are plenty of vegetables to pick, including delicious-looking eggplants and peppers. (Call first to find out what's available.) The farm also grows flowers in great abundance and has a large greenhouse filled with colorful plantings. These are spacious grounds, and you are welcome to walk through them.

29·

A RESTORED VILLAGE'S GARDENS

Vegetables and Herbs in Old Bethpage Village,
Old Bethpage

✻ DIRECTIONS

1303 Round Swamp Road, Old Bethpage, NY. Take the Long Island Expressway east to exit 48.
Take Round Swamp Road south, and look for the entrance almost immediately on your left.

✻ INFORMATION

Open March–December, Wednesday–Friday, 10 A.M.–4 P.M.; Saturday–Sunday, 10 A.M.–
5 P.M.; closed during January and February. Pick up a self-guided architectural walking
tour at the entrance, or take a guided tour. There is an admission fee. (Note: During
the school year, you are likely to find groups of schoolchildren on weekdays; if you
prefer to visit at a quieter time, plan accordingly.) Telephone: 516-572-8400. Web site:
http://www.nassaucountyny.gov/agencies/parks/WhereToGo/museums/central
_nass_museum/old_bethpage_rest.html.

✻ Old Bethpage Village Restoration is an outdoor museum of both historical and architectural interest. A re-creation of a mid-nineteenth-century Long
Island village, it includes more than sixty preserved historic structures picturesquely set amid two hundred acres of rolling hills. This collection of assorted
buildings—from carriage houses, shops, and cottages to rustic farm structures
—features simplified versions of Georgian, Greek Revival, Federal, and Gothic
styles that were found in Long Island's countryside from the mid-1700s to the
late 1800s.

The restoration was begun in 1962. Worthy historic structures that were
threatened by the region's rapidly growing suburban expansion were brought to
this site, once a vast farmland, and restored. Although the majority of these preserved buildings date to the mid-nineteenth century, there are earlier examples,
such as the eighteenth-century Shenck House. You are invited to walk from one
building to the next, where you will find costumed interpreters assuming the
roles of teachers, shopkeepers, blacksmiths, and farmers who typically lived
in this kind of rural community. Old Bethpage Village is still expanding, with
buildings of historical and architectural merit being added to the site; each is
carefully researched and restored and then furnished with authentic pieces.

As you walk from one antique building to the next, you'll find a variety of
vegetable and herb gardens, as well as the Powell Farm Barn. This English-
style farm complex dates to the 1850s. And behind several of the wonderful

old houses are the kitchen gardens (so called because they were for the use of the cook in the nearby kitchen, who presumably dashed out the kitchen door, cut some very fresh vegetables and herbs, and threw them into the pot on the stove!). These are not small gardens, by the way, being typically one to three acres; obviously the output was meant to feed a large family and, with canning and preservation, to last during the winter months as well.

In the spring and summer you'll find a nice variety of vegetables and herbs still being grown in four of these sites. Among the crops produced at Old Bethpage Village are cucumbers, carrots, leeks, tomatoes, lettuce, yellow squash, onions, and winter squash, as well as a variety of herbs.

We found that the sight of thriving vegetable gardens added greatly to the village's feeling of authenticity, suggesting how people actually lived and farmed in a typical mid-nineteenth-century Long Island community like this one.

AND KEEP IN MIND . . .

30 ◆ Quail Hill Farm, Amagansett

ADDRESS: *Old Stone Highway and Deep Lane, Amagansett, NY.*
INFORMATION: Telephone: 631-283-3195. Web site: http://www.peconiclandtrust.org/quail_hill_farm.html.

One of Long Island's longest standing CSAs (community-supported agricultural sites where people can join in farming their own plots) and one of the oldest organic farms in the state, this is an amazing place of some one hundred acres, where members can harvest their own vegetables, berries, and herbs. Operated by the Peconic Land Trust, it serves at least two hundred families, as well as local schools, food pantries, and nearby restaurants. Although Quail Hill Farm is a bit beyond our usual driving radius, it is worth a visit if you are on the East End and have an interest in community gardening. Call before going to make sure the farm is open.

31 ◆ Loughlin Vineyards, Sayville

ADDRESS: *100 South Main Street, Sayville, NY.*
INFORMATION: Telephone: 631-589-0027. Web site: www.loughlinvineyard.com.

Unusual to the South Shore of Long Island, this is a vineyard near the sea, which is reflected by the sandy paths between the rows of growing grapes. The six-acre site is closer to New York City than any other vineyard on Long Island and is easily accessible. Although Loughlin has not had a winery to process its grapes, it is completing one now. This is a pleasant place for a walk near the ocean.

NEW YORK STATE
East of the Hudson

32·
A CELEBRATION OF THE FARM
Stone Barns Center for Food and Agriculture, Pocantico Hills

✿ DIRECTIONS

630 Bedford Road, Pocantico Hills, NY. From Manhattan take the Henry Hudson Parkway to I-287 west, then Route 9 north, just before the Tappan Zee Bridge. Continue on Route 9, past Tarrytown, and take a right onto Route 448/Bedford Road. You will go up a steep hill; take a left at the stop sign, continuing on Route 448. Go about half a mile to the entrance, on your left.

✿ INFORMATION

Open year-round, Wednesday–Sunday, 10 A.M.–5 P.M. (closed on Monday and Tuesday). Telephone: 914-366-6200. Web site: www.stonebarnscenter.org.

✿ You drive along the shimmering Hudson, go up a steep hill, continue along a wooded road dotted with quietly elegant estates, and, after a few turns, come to a broad, pastoral landscape, reminiscent of the English countryside. You've reached Stone Barns. This storybook farm, complete with animals grazing in pastures, wood fences, and picturesque stone "barns," is much more than it seems at first glance. As stated in its informational materials, Stone Barns Center for Food and Agriculture (its official title) is "a farm, a kitchen, a classroom —an exhibit, a laboratory, a campus. The mission of this unique, nonprofit, member-driven collaboration is to celebrate, teach, and advance community-based food production and enjoyment, from farm to classroom to plate."

These words are not an overstatement: the one-of-a-kind center is—and does —many things. An organic working farm set on eighty idyllic acres, Stone Barns is dedicated to educating the public (children included) about locally raised food; it encourages visitors to explore the grounds (on their own or by guided tour) and to see at first hand how things are grown; it offers an impressive list

---

Farm-to-Table Experience

Stone Barns offers a rare experience for all of the family. You can spend a day farming in Stone Barns' fields (under supervision, of course) and then go back to the Resource Center, where the produce you've farmed and picked will be cooked up, and you get the chance to eat it! You are even given the recipe to take home.

Greenhouse at Stone Barns, Pocantico Hills (Courtesy Stone Barns, Anabel Braithwaite for Belathée Photography)

of programs for both children and adults: tastings, cooking classes, demonstrations, lectures, farmer-in-training (for kids), and how-to workshops. Stone Barns is also home to two restaurants: Blue Hill at Stone Barns (the more formal of the two) and Blue Hill Café (more suited to lunch or a snack), where freshly picked produce is always on the menu.

We can thank the Rockefeller family for the creation of Stone Barns Center for Food and Agriculture. In fact, the property was once their private farm, attached to the family estate in Pocantico Hills. The landscape was reconfigured in the 1920s under the expert eye of the Olmsted Brothers design firm, making it all the more appealing. A few years later, the Norman-style stone barns were built, adding a certain old-world elegance to the rustic setting. In the seventies, Peggy Rockefeller (wife of David Rockefeller) became personally involved in farming the land and raising cattle. After her death in the mid-1990s, David

Rockefeller donated the building and surrounding acres to the Stone Barns Restoration Corporation, in her memory. With funds from Rockefeller, this not-for-profit organization developed the property and restored the barns (something his wife had especially wanted). In 2004 the center as it is today was opened to the public.

From the moment you step onto the site, you know you're in for a rare treat. You first come to the stone barns and surrounding buildings, which form an attractive courtyard. There you'll see the visitor center, where you can get information and pick up a descriptive map of the farm for your self-guided tour. You are free to wander about at will and will find that things are well marked and the staff (and volunteers, of whom there are quite a few) both helpful and knowledgeable.

Because the farm is productive all year, with hundreds of varieties of plants growing both in the outdoor fields and gardens and in the greenhouse, there is always something to see. The diversity of produce is remarkable and includes rare items such as suiho, hakurei turnips, celtuse, and other names that are not familiar to most of us. As you would expect in a sustainable farm environment, pesticides, herbicides, and chemical additives are never used, and the suitability of crops for the local soil and climate is always taken into account.

For us, the highlight of the place is the immense (twenty-two-thousand-square-foot) greenhouse. This beautiful, light, and airy structure is filled with rows and rows of vegetables, tended with the utmost care. The greenhouse is

Beets at Stone Barns, Pocantico Hills (Courtesy Stone Barns, Anabel Braithwaite for Belathée Photography)

called an "in-soil" greenhouse and has a retractable roof to let in fresh air. We admired delicious-looking lettuces, spinach, radishes, kale, turnips, herbs, carrots, chard, and mache and walked up and down the aisles for a closer look and a couple of photos (no one seemed to mind).

If you're here with children, you'll want to walk past the farm animals (chickens, pigs, sheep, turkeys, geese, and bees). During the winter months (when there is no grass for them to graze on) the livestock are enclosed, but at other times of the year they can be found roaming through the pastures, where they graze freely. For the sheep this means grazing on the grass; for the chickens it means foraging for seeds and insects and exploring near and far; for the pigs it means rooting around in the shade of the forest. Visitors are encouraged to come at any time of the year to see animals in their natural habitat but are asked not to feed or disturb them.

Other sites not to miss include the Dooryard Garden, a demonstration garden focusing on growing techniques; the herb garden, where the restaurant chefs pick their herbs and edible flowers; and the Upper and Lower Vegetable Fields, the main outdoor growing area. There are areas (sometimes restricted) dedicated to livestock and their management, composting, and wetlands. On the outer edges of the property are scenic walking trails to Buttermilk Hill and to Swan Lake and Rockefeller State Park Preserve.

After all this exercise you'll want something to eat. The popular Blue Hill Restaurant, housed in a charming converted dairy barn with high, vaulted ceilings, is the reason many people come to Stone Barns in the first place! The more casual Blue Hill Café is a great place to stop for lunch or a snack. Here you can munch on freshly made salads or sandwiches, indoors or on the patio, looking out over the countryside. You can also check the farmers' market (located near the visitor center), which sells a variety of fresh produce three times a week from May through November, then once a month from December through April.

33 ·

THE RARITY OF AN AMERICAN CHESTNUT PLANTATION

Lasdon Park, Arboretum, and Veterans Memorial, Somers

ஜ DIRECTIONS

Route 35, Somers, NY. From I-684 take exit 6 and go west on Route 35. Follow signs for the park entrance, which is on Route 35, about three miles ahead on the left. From the Taconic State Parkway take the Route 202 exit and go east on Route 202; turn left at Route 35 and go about three miles. The park entrance is on your right. Follow signs to parking area.

ஜ INFORMATION

Open year-round, daily, 8 A.M.–4 P.M. Admission is free. The chestnut plantation is quite a long walk from the main parking lot; to see it up close and avoid extra walking, you can drive to another, much smaller parking area next to the entrance to the Chinese Friendship Pavilion and Cultural Garden, following these directions: turn right at the imposing entrance gates (instead of left, which leads to the main parking area) and follow the road, bearing left. You will see the small area for cars next to the sign for the Chinese Friendship Pavilion; from there you walk down a grassy slope to a broad field; you will see the chestnut grove on your left, enclosed within a fenced area. Telephone: 914-864-7263. Web site: www.westchester.com.

ஜ Lasdon Park, Arboretum, and Veterans Memorial is one of the best-kept secrets in the region. Not nearly as well known as you might expect, given the many offerings here, this spectacular site—reminiscent of an English estate with a manor, formal gardens, broad lawns and meadows, magnificent trees, woodlands, and vistas everywhere—is a not-to-be-missed pleasure. But there is something else here that makes it extra special: within the property is a chestnut plantation, a true rarity these days.

Until the early twentieth century, American chestnut trees were widespread throughout the eastern part of the country. A valuable economic resource, their timber was used to build barns and houses, and their nuts fed both humans and wildlife. Unfortunately, a fungus was accidentally imported between 1901 and 1908, causing a devastating blight. By 1940 most (over three billion) American chestnut trees were completely destroyed. Over the years organizations such as the American Chestnut Foundation have been working diligently to breed a blight-resistant American chestnut and to restore it to forests of the East. For now, the American chestnut trees here at Lasdon are among the few examples of this rare breed to be found in the area.

There is a lot to see within the 234 acres that make up Lasdon Park (and quite a bit of walking), and you can decide whether to start with the chestnut plantation (in which case you would follow the parking directions given in the Information section) or the park/arboretum, accessed by the main parking lot. A general overview of the park is probably best, as it gives you a wonderful sense of the place and its vastness. At the entrance kiosk, pick up a guide to the park (which includes a handy map of the park's many features). First you come to a memorial garden honoring William and Mildred Lasdon, who owned the property from 1940 until it was sold to the county in 1986. (Their bronze busts are displayed here, amid flowers and shrubs in a pretty courtyard.) The Lasdons were horticulture enthusiasts and imported many of the tree specimens found in the thirty-acre arboretum. Just beyond the entrance you'll see a formal garden with boxwood hedges, a central fountain, and flowering bulbs, followed by a garden of shrubs (some with edible berries) bearing names starting with every letter in the alphabet, from A to Z.

Follow the signs to the sites that interest you most, as this is a large area. But if you like to walk, you may want to see it all! Everything is well marked and properly labeled (unlike some arboretums we have visited in our forays). A highlight is the beautiful street tree grove, followed by the flowering tree grove and apple orchard. Fruit trees along this inviting alley include peaches, pears, and nectarines. The arboretum also has impressive pine and lilac collections, an azalea garden, and a yellow magnolia grove. The so-called Famous and Historic Tree Trail features species connected with well-known Americans. (You can read about each example at a series of stations.) A bit of a walk will take you to the main house (dating from 1933 and now used as the office), a colonial mansion fashioned after Mount Vernon.

When you have finished exploring this section of the estate, take your car and drive around to the smaller parking area located near the entrance to the Chinese Friendship Pavilion and Cultural Garden. If time allows, walk down the marked path that leads to this pretty pavilion. Set within a Chinese-style garden with bamboo and kousa dogwood, it sits serenely on the edge of a pond. The pavilion was a gift from China to the people of Westchester.

The access to the chestnut plantation is to the right of the entrance to the Chinese pavilion. There is no formal path leading to the plantation, so you have to make your own way down the grassy slope, which is easy enough; as you approach, you will see it before you. Don't expect tall, mature trees; these are still quite young, having been planted not too long ago. The fact that they exist at all—and are being carefully monitored—is encouraging for the future of this nearly vanished species. There is something quite exciting about seeing at first hand an experiment of this kind and in such a spectacular setting, too! Few people come to this spot (except for the occasional staff member), and this is about as peaceful a place as you will find anywhere, as well as a hopeful sign that an American classic can be revived.

34·
HANDS-ON

Maple Sugaring and Apple Cidering at the Greenburgh Nature Center, Scarsdale

✍ DIRECTIONS

99 Dromore Avenue, Scarsdale, NY. Take the New York State Thruway (I-87), to exit 5 (Central Avenue or Route 100). Turn right and continue for five miles. Shortly after you see Scarsdale Woods, make a right turn on Dromore Road and follow signs to the center.

✍ INFORMATION

Open year-round, dawn to dusk. There are many educational and hands-on programs offered. Call for information on maple sugaring or apple cider making. Telephone: 914-723-3470. Web site: www.greenburghnaturecenter.org.

✍ This is an unusually interactive nature center of some thirty-three acres, where adults and children can all enjoy programs ranging from aquatic and avian adventures to making maple sugar or apple cider. On a pretty site with a lovely old manor house and all kinds of natural scenery—ponds, forest, greenhouse, and so on—a variety of experiences are offered here. There are also a live animal and a birds-of-prey collection and a variety of sites for studying plant ecology, biology, and related subjects.

The apple cidering is an autumn activity; children can use an old-fashioned screw press to create cider. In late winter/early spring (depending on the weather) the maple-sugaring events take place at three different locations within the Nature Center.

The first event is a visit to the sugar bush area to see how maples are tapped for sap. Both Native American and colonial techniques of making the syrup from the sap are demonstrated at sugaring sites. (A small sugar shack is used to boil the syrup.)

A Few Words on Maple Sugaring

Maple sugar/syrup, first used by Native Americans, has for years been an important American staple. It's not too difficult to make, though the process is fairly slow, and there are many places throughout our region where you can watch it in progress. (Or you can try to make it yourself!)

The best trees to tap are sugar and red maples, both common in the Northeast. (Trees must be of a certain diameter to sustain tapping, with only one to three taps maximum per tree.) If you figure that a single taphole normally yields about five to ten gallons a season (sometimes more, if you're lucky) and that it takes about ten gallons of sap to produce one quart of syrup, you can see why the stuff is so expensive!

Usually you can observe maple sugaring during February and early March (depending on the weather). Here is how it works: in late summer and fall, as maple trees temporarily stop growing, they begin to store excess starches, a process that continues throughout the cold season. As temperatures rise, the starch is converted to sugar, which passes into the tree sap. Cold nights with warmer days create pressure inside the tree, causing sap to flow. Trees are tapped, and the sap drips out. Buckets or tubes, which have been hung on the hooks of spouts, are filled with the precious liquid, which is then boiled and processed through various stages to become maple syrup. And voilà!

35 ·
RE-CREATING THE PAST
WITH PLANTINGS

A *Slave Garden on the Colonial Estate of*
Philipsburg Manor, Sleepy Hollow

✿ DIRECTIONS

381 North Broadway, Sleepy Hollow, NY. From New York City take the Henry Hudson Parkway to Route 9 north, to Tarrytown. The site is two miles north of the Tappan Zee Bridge.

✿ INFORMATION

Open April–December, daily except Tuesday, 10 A.M.–5 P.M. In March it is open only on weekends, and it is closed during January and February, as well as on Thanksgiving and Christmas Day. Demonstrations are offered, as are hands-on tours for children (on weekends). Telephone: 914-631-8200. Web site: www.hudsonvalley.org/content/view/14/44/.

✿ Philipsburg Manor is among the most picturesque sites you'll find in the Hudson Valley. This rural property was once a vast plantation dating back to the late seventeenth century; now, much reduced, it is centered around a

Slave Garden, Philipsburg Manor, Sleepy Hollow (Courtesy Historic Hudson Valley)

restored manor house and working gristmill. Of particular interest to us are the gardens: with varieties cultivated more than two hundred years ago, vegetable gardens have been created that replicate the past. The most evocative is the Slave Garden.

Once a bustling place with a large population of farmhands (including enslaved Africans), Philipsburg Manor was the first industrial complex in the thirteen original colonies. At its height, the mill produced so much flour and meal that much of it was shipped overseas. The ninety-thousand-acre property included not only the present manor but also the private residence of the Philipse family (located in what is now Yonkers); the Philipse family operated this flourishing enterprise until after the Revolutionary War, when, because of their Loyalist sentiments, they were forced to flee to London. Many years later and in disrepair, the manor was purchased by the Rockefeller family, restored, and eventually opened to the public. It is now operated by Historic Hudson Valley, along with several other nearby sites.

You can visit the house and mill on your own, following a self-guided tour. The whitewashed fieldstone manor house with its gambrel roof is relatively small, though it is double the original size (it was added to in the 1750s). It was set up more as a place of business than an actual home, with space for work, office, living, and storage. Most of the furnishings (some are authentic to the period, though not necessarily original to the site, and others are reproductions) are fine eighteenth-century examples of Dutch and regional pieces.

The highlight at Philipsburg is certainly the mill, everyone's favorite (including children, who also are likely to be intrigued by the costumed miller's demonstrations of its workings). The romantic image of the mill with its wooden waterwheel, reminiscent of a Currier & Ives print, is one that is most captivating to visitors.

But the pleasures of colonial manor life are offset by your visit to the Slave Garden, where you see the underside of the prosperous colonial enterprise. In this re-creation you see a kitchen garden such as African slaves would have cultivated. There are black-eyed peas, okra, cayenne peppers, sweet potatoes—all era-appropriate choices for the slaves' own use. In addition, one section of the garden is devoted to vegetables that the owners could have sold to northern European colonists.

This is an educational and evocative visit, particularly for families with children in tow. The entire complex—manor house, mill, and gardens—gives an unusually clear visual picture of life in the colonial-era Hudson Valley.

36·
FRUIT PICKING IN THE LOWER HUDSON VALLEY

Wilkens Fruit and Fir Farm, Yorktown Heights;
and Stuart's Fruit Farm, Granite Springs

☙ DIRECTIONS

Wilkens Fruit and Fir Farm: *1335 White Hill Road, Yorktown Heights, NY. Take the Taconic State Parkway to the Route 202/Yorktown Heights exit; go west on Route 202/35 for about one hundred yards and turn right at the traffic light onto Mohansic Avenue; go for about one mile to the end of the road and turn right at White Hill Road. The farm entrance is at the top of the hill.*

Stuart's Fruit Farm: *62 Granite Springs Road, Granite Springs, NY. Take the Taconic State Parkway to the exit at Underhill Avenue; at the exit make a right and then a left at the first traffic light (Route 118). Continue on Route 118 and go through two traffic lights; at the third traffic light take a left onto Mahopac Avenue. Continue on Mahopac Avenue until the first stop sign and make a left. You'll find the farm a quarter mile up the road on the right.*

☙ INFORMATION

Wilkens Fruit and Fir Farm: Open August–December, daily, 10 A.M.–5 P.M.; the pick-your-own section of the farm is open Friday–Sunday, 10 A.M.–5 P.M. Telephone: 914-245-5111. Web site: www.wilkensfarm.com.

Stuart's Farm: Open May–Fourth of July, 9 A.M.–6 P.M.; August–Christmas, 9 A.M.–6 P.M. Telephone: 914-245-2784. Web site: www.stuartsfarm.com.

☙ In this region there are fewer farms these days than there used to be—mostly for economic reasons—but many of those that remain have opened sections of their orchards and fields to the public to help their business. If you're looking for a pleasant way to spend a summer or fall day in the scenic Hudson Valley, consider heading to one or two of these pick-your-own farms to experience something a bit different. We recommend these two farms/orchards, which have the benefit of being close to each other and are thus easily explored in just a few hours. Combining the beautiful scenery of Westchester's rolling hills with an entertaining activity for all ages, both are appealing destinations, well worth a trip. And be sure to bring the children, as they will certainly enjoy the tasty adventure of picking their own fruit!

The first, Wilkens Farm, is a breathtaking site, with lovely old trees and winding pathways. Picturesquely set atop a steep hill, with spectacular views all around, the vast grounds include 180 acres of orchards, fields, and walking trails. This is the perfect place to explore and enjoy nature's bounty at first hand.

Grandma's Peaches with Raspberry Sauce

Serves 6
2¼ cups sugar
3 cups water
1½ teaspoon vanilla extract
6 fresh peaches
2 packages frozen raspberries, thawed and drained

In a heavy pan over medium heat stir together 1½ cups sugar, water, and vanilla until sugar is dissolved. Add whole, unpeeled peaches and simmer nine minutes. Remove pan from heat and allow peaches to cool in syrup thirty minutes. Drain and peel. Chill.

Puree the raspberries with ¾ cup sugar in a blender until the sugar is completely dissolved. Chill.

Before serving, cut peaches in half and arrange on plates with raspberry sauce poured over them.

This is an almost perfect summer dessert, refreshing but not filling. Certainly it is an ideal complement to a substantial meal.

Though this pick-your-own farm specializes in apples (at least thirty varieties are grown here, half of which are available for picking), you can also gather peaches, nectarines, and pumpkins. In addition, the farm grows good-sized fir trees—five to ten feet tall; you can choose, then cut down, your own Christmas tree.

If you're here primarily to pick fruit, stop first at the Rainbow Pavilion to get your baskets and picking poles before walking (or riding on an old-fashioned hay wagon) to the appropriate orchard or field. Peaches and nectarines are ripe in late summer, while apple picking begins in the first week of September (when Gala apples begin to ripen) and goes through at least the last week of October, with the appearance of Granny Smiths, Fujis, and Idareds. If you would like to take home apples or other fruit without the effort of picking your own, stop at the well-stocked farm market (located in a bright-red barn), which offers quite an array of locally grown produce. We also recommend the pure natural apple cider, which is freshly pressed in the farm's own mill (worth a visit).

Our other site, Stuart's Fruit Farm, is a vast two-hundred-acre property. Though at first it appears to be somewhat stark, with few trees in front surrounding the buildings, it does have acres of fields and orchards behind, where you can enjoy walking—and picking. Stuart's also offers an impressive variety of apples for picking, as well as peaches and pumpkins. Gather as many as you can carry, then check out the farm market for the blueberries, plums, squash, broccoli, tomatoes, cucumbers, lettuce, cauliflower, onions, and more. Like Wilkens, Stuart's offers such favorites as honey (there is a beehive on the premises), cider, jams, syrup, and eggs.

If you're a history buff, you'll be interested to know that Stuart's Fruit Farm has the distinction of being the oldest working farm in Westchester. Indeed, since 1828 it has been owned and run by the Stuart family and is considered to be a landmark in the community. The farmhouse (still occupied) dates from 1760, and the fruit stand from 1886. You'll find the folks here unusually friendly and ready to answer questions or offer suggestions about planting or picking.

So You Think You Would Like to Farm?

Farming is becoming increasingly popular as a profession among the young. But before you get into it you might take a look into a special site where you can try it out. Common Ground Farm in Wappingers Falls, New York, describes itself as a "kind of grad school for farmers," who can experience everything from hands-on labor to discussions of sustainable agriculture. The nonprofit organization sponsors educational programs and offers shares to low-income members, as well as apprenticeships to would-be farmers, who can discover what farming is really like. Web site: www.commongroundfarm.org.

37·
DUTCH COLONIAL HERBS AND VEGETABLES

Van Cortlandt Manor's Period Authentic Gardens, Croton-on-Hudson

᪥ DIRECTIONS

500 South Riverside Avenue, Croton-on-Hudson, NY. Take the New York State Thruway (I-87) to exit 9; go north on Route 9 past Ossining to Croton Point Avenue. Turn right and go one block to Riverside Avenue. Turn right again and go a quarter mile to Van Cortlandt Manor.

᪥ INFORMATION

Open daily (except Tuesdays and major holidays), 10 A.M.–5 P.M. There is an admission fee. Tours are available. Telephone: 914-271-8981. Web site: www.hudsonvalley.org/content/view/15/45/.

᪥ For those of our readers who enjoy colonial history, as well as period-authentic gardens, Van Cortlandt Manor is a good place to visit. In addition to the symmetrical, white-porched Dutch patent house, which dates to the 1680s, there are reconstructed historic gardens and many interesting events to enjoy.

Van Cortlandt Manor, Croton-on-Hudson (Courtesy Historic Hudson Valley)

The eighteenth-century garden, which is part of the twenty-acre estate, includes herbs and vegetables, in particular the kinds of foods that colonial families would have grown: potatoes, squash, beans, and so on. (There are no tomatoes, which they considered poisonous.) There are over twenty herbs known to have been grown at that time, and the settings are typical of working gardens of the Dutch colonial era.

While you are there, of course, we recommend a visit to the manor house itself—a fascinating place of history and charm.

38 ·

FROM QUINCE TO GINGKO
TO BAMBOO SHOOTS

The Hammond Museum and Japanese Stroll Garden's
Wonderful Collection, North Salem

❧ DIRECTIONS

28 Deveau Road, North Salem, NY. Take I-684 north to exit 7 (Purdys/Somers); go right at Route
116 east and then left at Route 116 east/Route 22 north; continue on Route 116 east and make a
left at June Road; take the first right onto Deveau Road.

❧ INFORMATION

Open April–end of November, Wednesday–Saturday, noon–4 P.M. There is an admission fee. Telephone: 914-669-5033. Web site: www.hammondmuseum.org.

❧ This intimate Japanese garden is situated at the end of a rural road in the northeast corner of Westchester County. Befitting the idea of an Asian garden as an oasis for contemplation, the site is about as peaceful a place as you could find. A small gravel path dotted with stepping stones winds around a lotus pond and past groves of specimen trees, shrubs, and tiny individual gardens. With trail guide in hand (available at the entrance) you can identify katsura trees, bamboos, Chinese chestnuts, crabapples, larches, locusts, cedars, and many others in a long list. There are edible varieties, too: pear and cherry trees, Japanese flowering quince, highbush blueberries, gingkos (used as an herbal supplement), and bamboo shoots. Carefully placed rocks and stone statues are scattered here and there: on a small island in the pond, by a flowering shrub, next to raked gravel gardens or the gently cascading miniature waterfall.

A visit to this meditative site is a sensual experience: you can hear the sounds of songbirds, crickets, frogs, or breezes blowing through the bamboo; smell the delicate aroma of the plants; feast your eyes on the inspiring gardens; feel the soft touch of leaves as you walk along—and simply imagine the delicious taste of the fruit you find amid the greenery. This is nature's bounty in its most subtle mode. You'll find garden benches along the way from which to contemplate it all.

The garden is a delight at any time, and especially so in springtime, with blooming cherry trees, azaleas, and irises, or in fall with its vibrant colors. A little outdoor café beneath a grove of plane trees serves a pleasant lunch during the summer.

Gingko tree (iStockphoto.com/Sondra Paulson)

39·
PEACHES, PEARS, CHERRIES, AND APPLES ON A HILLSIDE

Pick-Your-Own at Fishkill Farms, Hopewell Junction

✍ DIRECTIONS

9 Fishkill Farm Road, Hopewell Junction, NY. Take the New York State Thruway (I-87) to I-84 east, to exit 15. Turn left at end of the ramp; you will be on Lime Kiln Road. Go less than a mile to the stop sign. Turn right on East Hook Cross Road. Fishkill Farm Road is on your right.

✍ INFORMATION

Fruit ready to pick from late June to October. Telephone: 845-897-4377. Web site: www.fishkillfarms.com.

✍ This is the kind of panoramic farm vista that we like to imagine, with rolling hills dotted with orderly rows of fruit trees stretching out as far as you can see. You arrive at the summit of a hill with fields sloping down around the barn and parking area. During picking season you walk down the hillside among the trees—a beautiful walk even if you leave empty-handed. (Produce is also sold on the site, already picked.) There are some forty varieties of apples, as well as peaches, cherries, nectarines, and pears. There is also a big vegetable garden and berry patches (blueberries, blackberries, and currants).

We enjoyed the color-coded map (available at the Farm Market near the barn) that lists the wonderful variety of apples and when they are ready ("Early Gold in late August, Macoun in mid-September, Jonagold in October, McIntosh, Mutsu," etc.). Long stretches of bright colors on the map show visitors where to go to find their favorites.

This is a historic place; Fishkill Farms is over one hundred years old. They say they are "committed to growing food in harmony with nature" and sustainable agricultural practices, including ecological alternatives to conventional

---

How to Prepare Your Own Small Garden

Muscoot Farm in Westchester offers a class in hands-on gardening, preparation, and planting on a regular basis in its historic garden. Call ahead for information: 914-864-7282.

pesticides. (The apples are sprayed with kaolin, a fine nontoxic form of clay, to protect them from insects.)

This venerable farm/orchard is near enough to the urban area that you can easily visit (and pick) on an afternoon outing with your family. (There are strict rules against climbing up the trees, but picking poles for out-of-reach fruit are provided.)

40·
EDIBLE EXOTICS IN A BOTANICAL PRESERVE

Enjoying the Wilderness at the Mianus River Gorge
Wildlife Refuge and Botanical Preserve, Bedford

ஃ DIRECTIONS

167 Mianus River Road, Bedford, NY. Take the New York State Thruway (I-87) north to the Cross County Parkway. Go east to the Hutchinson River Parkway to I-684 north. Exit at Route 22 north and follow the signs for Route 172 east. Turn right at Long Ridge Road (where you will also see a sign for Stamford, Connecticut), continue for about a mile, turn right on Millers Mill Road, and then turn left on Mianus River Road. The gorge is half a mile down the road on your left.

ஃ INFORMATION

Open April–November, daily, 8:30 A.M.–5 P.M. Call for information or to request help locating the edible exotics. Telephone: 914-234-3455. Web site: www.mianus.org.

ஃ An excursion to the Mianus River Gorge Wildlife Refuge and Botanical Preserve is ideal if you are seeking an unspoiled wilderness in relatively close proximity to New York City. Located less than forty miles northeast of Manhattan in a rural section of Westchester County, the park includes about 450 acres of pristine woodland, mostly hemlock and beech, along the banks and gorge of the Mianus River. The gorge is a deep ravine that was carved out by a great sheet of ice some ten thousand or more years ago. The name of the river comes from Myanos, a local Indian chief the white settlers met when they first arrived in the seventeenth century.

This is an inviting spot for a quiet walk in a natural and tranquil sylvan setting. The only sounds you hear are those of birds (there are more than 150 species) or the rushing river in the ravine or a nearby gurgling brook. There are five miles of foot trails that go up and down over hilly terrain, amid a rich variety of plant and animal life. The dense, dark evergreen forest alternates with leafy, airy woodlands of maples, beeches, tulip trees, and dogwoods, so that you go in and out of shadows, with varying degrees of sunlight filtering through the foliage.

Of particular interest to us were what the chief plant specialist termed "edible exotics." Among these unusual growing edibles (which you may not pick, of course) is garlic mustard, whose sharp notched leaves and little white flowers in spring identify it. A type of herb, the leaves were used by early American colonists.

Another edible exotic is knotweed, also called Japanese bamboo, which is eaten like asparagus. A third edible exotic you'll find growing in this beautiful spot is wineberry, a relative of the raspberry, used to make fruit wine. Wineberries show up on three- or four-foot-tall bushes during the second or third week of August.

Wild ginger has also been spotted at Mianus, along with various mushrooms (not for picking or eating). A walk here is an entirely different experience from the more traditional growing venues. It is wonderful for urban dwellers who seldom have the opportunity to experience nature firsthand. We recommend it also for adventuresome children who love to explore and who should be intrigued by an enormous hemlock reputed to be more than three hundred years old. Although the trails are quite hilly, they are not too difficult to manage if you proceed at a leisurely pace. (However, we do not recommend them for those who find climbing arduous.)

The approach to the preserve sets the tone for what is to come. From Miller's Mill Road you drive along Miller's Mill Dam with its impressive cascade, bearing to the left and following discreet signs to the entrance of the preserve. Park your car, and at the shelter pick up a trail map, which gives you a good description of where you are going. Basically you will be staying on one trail, which is marked in red going out and blue coming back. There are detours worth making along the way, all of which are indicated on the trail guide.

As you start on your walk, you are struck by the silence around you. After a moment you will see a small wooden bench on your right, slightly off the trail, with a view overlooking the beginning of the gorge. This bench, called the Lucy D. S. Adams Memorial Bench, was placed there for those who cannot walk the trails but want to enjoy this special environment.

You proceed over a soft carpet of pine needles, leaves, and mosses, always following the well-marked trail. You cross a charming little stone bridge over a brook with masses of ferns and beeches alongside. The path leads you up and down hills, with occasional views of the gorge below. About a half hour into the walk you will come to a place where you make a sharp turn up a rather steep hill with random stone steps and gnarled tree roots. You are now on your way to the "Hemlock Cathedral," a twenty-acre site of tall, majestic trees atop a hill. This is a quiet place. The sunlight filtering through the trees creates the impression of the interior of a Gothic church. You then pass a beech grove and a wild ginger colony and will see a series of lovely old stone walls along the way that marked the boundaries between farms of long ago. A recommended detour is Hobby Hill Quarry, where mica, quartz, and feldspar were mined during the eighteenth century. We also suggest you take a few minutes and follow the signs for the reservoir view.

For those interested in finding the "edible exotics" mentioned earlier, we recommend calling ahead and asking to speak with the chief garden expert, who offered to show us these plants.

41 ·
A SPECTACULAR GARDEN
WITH A RIVER VIEW

Water, Rocks, Flowers, Vegetables, and Herbs at
Stonecrop Gardens, Cold Spring

❧ DIRECTIONS

81 Stonecrop Lane, Cold Spring, NY. Take the Taconic State Parkway to the exit at Route 301.
Go west for about three miles. The driveway for Stonecrop is a sharp right, directly opposite Den-
nytown Road (the sign for Stonecrop is very small and easy to miss).

❧ INFORMATION

Open April–October, Monday–Thursday, 10 A.M.–5 P.M.; Friday, 10 A.M.–dusk; weekends, check before going. There is an admission fee. Telephone: 845-265-2000. Web
site: www.stonecrop.org.

❧ Anyone who likes gardens should find Stonecrop Gardens inspiring. For
rock garden enthusiasts, however, these gardens are a must. A steep hill off
a rural road in Putnam County's rolling countryside leads to this enchanting
spot. Here, in an unusually idyllic setting, are some of the most glorious alpine
and water gardens anywhere.

The Stonecrop estate enjoys pastoral views over meadows to distant hills.
The beautifully designed grounds include a French-style country house with
adjoining stable and wood fences, an enclosed garden, potting sheds and greenhouses, ponds and a lake, stone walls, and the gardens. A woodland garden,
pond garden, grass garden, vegetable garden, herb garden, and perennial borders are among the many tasteful plantings.

You'll find rock gardens throughout the grounds—from the area next to the
house, where the plantings are displayed in tidy beds and in greenhouses, to the
magnificent stream and cliff rock gardens beyond. And it's likely that you won't
meet more than a handful of other visitors; you may have the place almost to
yourself.

Before embarking on your exploration of Stonecrop, stop at the office (located just inside the house) to pay the entrance fee and pick up a map and descriptive guide. You might start with the enclosed garden, accessible from a
scenic deck on the side of the house. (The panoramic view from here is spectacular.) Within a high wooden fence is an English-style garden with square and
triangular beds containing vegetables and old-fashioned flowers.

In fact, there are many great things for the table that are grown at Stonecrop. Among the thriving vegetables are parsnips, runner beans, several kinds of

chard, lettuce, and beautiful tomatoes. Not far away are herbs: rosemary, lemon verbena, and basil, among others. You'll find these treats near the statue of the patron saint of so many American gardens: Gertrude Jekyll (who resembles a scarecrow here). You can walk around winding pathways to see espaliers of dwarf apples and pears, as well as a lovely grape arbor.

A path leads to the greenhouses and raised glass-covered troughs for displaying alpines. The carefully labeled exhibits are of museum quality and include every imaginable variety, most shown in the tiniest of pots in neat rows. If you are interested in learning in detail about these plants or care to embellish your own rock garden, this is a perfect opportunity. (You might compare notes with one of the gardeners, who can often be found working diligently in this area.)

You'll see other examples of rock gardens in front of the house. Many are on raised beds supported by stone or limestone (tufa) walls; especially delightful are miniversions shown on rectangular and round "pedestals."

To reach the most spectacular site of all—the rock ledge and the stream garden that precedes it—walk west from the house. What you see here more dramatically than anywhere else at Stonecrop is the result of an imaginative partnership between nature and human ingenuity. In the 1980s the naturally rocky terrain was enhanced by the addition of yet more rocks, including giant boulders, on the ledge. A network of gently flowing streams and pools was created, emptying into a lake below (with the water recirculating through underground pipes), and thousands of plants—mostly alpines, grasses, dwarf conifers, and Mediterranean species—were carefully placed for color, texture, and pattern.

The visual effect of the streams gently moving through the delicate plantings, around the rounded rocks, and into the clear pools is magical. Best of all is the fact that you actually walk down the cliff garden, stepping onto rocks that form it; in so doing, you feel more like a participant than a passive observer. As with the rest of Stonecrop, everything here is beautifully maintained, and the plants are all labeled, even the tiniest. You can wend your way on a path of stepping-stones at water's edge to a charming wood pavilion covered with wisteria and similar in design to the main house (although in a more Japanese vein); this is a good lookout from which to enjoy the view.

The network of paths continues: around the lake, across a rustic stone bridge (known as the Flintstone bridge), and down the hillside toward the woodland pond. This lower pond is surrounded by primroses and woodland plants; from it a path leads through a grove of bamboo. You can explore it all at will, consulting your map.

Before leaving Stonecrop, be sure to walk on a small path through the woodland garden: azaleas, rhododendrons, and other shade-loving plants have been carefully placed to blend harmoniously with this natural habitat. Nearby is a pond surrounded by lilies and groupings of an exotic species with giant leaves (apparently the largest herbaceous plant recognized). You would hardly imagine that this delightful pond was once a swamp.

42·
AN ORANGERIE AND OTHER DELIGHTS

The Historic Estate of Boscobel, Garrison

℅ DIRECTIONS

1601 Route 9D, Garrison, NY. From New York City take the Henry Hudson Parkway, which becomes the Saw Mill River Parkway, as far as the Bear Mountain Bridge. Do not cross the river but pick up Route 9D and go north toward Cold Spring. Boscobel's entrance is on your left just after Garrison.

℅ INFORMATION

Open April–October, 9:30 A.M.–5 P.M.; check for winter hours. There is an admission fee. Telephone: 845-265-3638. Web site: www.boscobel.org.

℅ The garden at Boscobel—an elegant, historic estate high above the Hudson River—might be described as a garden with a view. We were entranced by the combination of artistic perfection—for such is the arrangement of Boscobel's dainty gardens—and the wild and natural patterns of the river below.

Boscobel has a fine house in the Federal style (which you may visit to see English and American antiques and paintings) set majestically on a wide expanse of sloping lawn. The house was built in 1806 by States Dyckman following a pattern of the great Robert Adam. The estate includes a series of gardens nearer the bluff overlooking the river. The grounds—some thirty acres—are varied and well worth walking through.

Of particular interest to us were the edible specialties: a real orangerie with orange, grapefruit, lemon, and lime trees in its glassed interior; you don't have to journey to Florida to see these beautiful, thriving fruit trees! There is also an apple orchard (which helps to support Boscobel), fig trees, an herb garden, and old-fashioned beehives (called skeps).

These flourishing sections of Boscobel are integral parts of the exquisite gardens of flowers. Tubs of oleander line the interlocking walkways, for example, and boxwood edges the gardens with geometric design. From almost every spot you may glimpse the majestic Hudson River below.

Although Boscobel is not unknown to tourists, it is generally not a crowded place, especially during the week. There are some of the trappings of success, as at all restorations, but we found the site quite unspoiled and, in fact, a place for both aesthetic pleasure and quiet contemplation.

43·
FROM CHINESE VEGETABLES
TO EGYPTIAN ONIONS

*Ryder Farm, a Historic Farm with a
"Certain Mystique," Brewster*

✍ DIRECTIONS

404 Star Ridge Road, Brewster, NY. Take the New York State Thruway (I-87) to I-84 east, to its intersection with I-684, to exit 9. Take Hard Scrabble Road east (right) to its end at a Y; turn left on June Road. Go a quarter mile (you'll see the high school on your left) and turn right on Bloomer Road. Go one hundred yards to Star Ridge Road. The farm is two-thirds of a mile ahead.

✍ INFORMATION

Open daily during growing season. We recommend calling before you visit. You can see a video on the farm's Web site. Telephone: 845-279-3984. Web site: www.ryder farmorganic.com.

✍ Ryder Farm has been in the same family since the 1790s. Covering some 130 acres, it is one of those evocative places that actually suggest another time and place—though its organic concepts are thoroughly up-to-date. There are old wagon roads here, vines reaching over the barns and sheds, old tractors, great trees, and an antique homestead—and in the midst of this cluttered, lovely scene, rows and rows of vegetables. (In fact, this is also a CSA farm, which means that outsiders can opt to grow their own produce in various spots.) But the place nevertheless suggests the family farm. This is the kind of setting you might want to walk all the way around, absorbing the sights and smells: the sweet onions and garlic growing, the rows of Chinese vegetables. In fact, as a most welcoming owner, Hall Gibson, described it, the place has a "certain mystique."

High on a ridge and sloping downward to Peach Lake, the land was settled by the Ryder family (arriving with oxen and chickens in 1795). The great-great-great-grandchildren are still farming it today. Ancient pathways and wagon trails attest to its long history.

But despite the farm's venerable past, the farmers here are interested in new ideas. There are forty-inch planting beds that are separated by grassed sod (to combat soil erosion), and the traffic of farmers and tractors is solely on the grassy strips. (This is a gardening concept of an influential agricultural thinker named Lee Reich, whose book is called *Weedless Gardening*.)

The organically grown vegetable list is long and inclusive; as the farmer said, it ranges from "arugula to zucchini." Of particular interest are the Chinese vegetables (including tatsoi, a form of mustard greens; mizuna; and bok choi); the many herbs (among them thyme, lovage, dill, catnip, and cilantro); the wide variety of tomatoes with their unusual names (Early Girl, Hogs Heart, Brandywine); Egyptian onions, chards, chives, and even raspberries (which you can pick yourself).

This is the kind of place where people take the time to talk with you—not a giant agricultural and commercial establishment. We hated to leave the "mystique" of Ryder Farm.

44 •

TWO IDYLLIC SITES IN
THE HUDSON VALLEY

Clinton Vineyards, Clinton Corners; and
Millbrook Vineyards and Winery, Millbrook

❧ DIRECTIONS

Clinton Vineyards: *450 Schultzville Road, Clinton Corners, NY. Take the Taconic State Parkway to the Salt Point Turnpike/Route 115 exit; go east on Salt Point Turnpike, through Clinton Corners, and go left on Clinton Corners Schultzville Road. Turn left onto Harpers Road.*

Millbrook Vineyards and Winery: *26 Wing Road, Millbrook, NY. Take the Taconic State Parkway to the exit at Route 44/Millbrook. Follow Route 44 east for about one mile, then take Route 82 north for three miles, to Route 57. Turn right on Route 57 and go three miles, turning left onto Wing Road. Vineyard is on your right.*

❧ INFORMATION

Clinton Vineyards: Winery and tasting room open year-round, Thursday–Monday, noon–6 P.M. (in winter, Friday–Sunday, noon–4:30 P.M.). Other times are available by appointment. Telephone: 845-266-5372. Web site: www.clintonvineyards.com.

Millbrook Vineyards and Winery: Open year-round, daily, noon–5 P.M., for guided tours and wine tastings. From Memorial Day to Labor Day, open 11 A.M.–6 P.M. Telephone: 845-677-8383. Web site: www.millbrookwine.com.

❧ While the number of farms has been dwindling in the Hudson Valley, the list of vineyards is booming, as more people are discovering the pleasures of visiting them and tasting their offerings. These wineries range from simple, rustic sites to glamorous venues that regularly host elegant events. Two of our favorites, near each other, are Clinton and Millbrook vineyards, beautifully situated in the gentle hills of Dutchess County. Though their style and look are quite different, both offer a welcoming ambiance, spectacular scenery, pleasant walking, and a good variety of delicious wines to taste and buy. A day's trip to visit them is a real treat—especially during harvest time, when you can witness firsthand the fascinating process and hard work involved in producing wine.

Clinton Vineyards is the more rustic and intimate of the two. Called the jewel of Hudson Valley agriculture, this idyllic site of one hundred acres includes historic Dutch barns, gardens, a pond, fields, woods, and of course the vineyard itself, which stretches out over a sunny slope above the winery. The beauty of the landscape invites exploration, and you are more than welcome to walk along winding paths through the vineyard and nearby fields.

Millbrook Vineyards, Millbrook (Courtesy Millbrook Vineyards & Winery)

Everything is appealing and low-key about this place, from the naturalistic landscape itself to the unpretentious and charming tasting room (with its glorious views of the vineyard) and the warm welcome you'll receive from the proprietor and friendly staff (all of whom are happy to chat and give out information as needed). And of course the wines are very special, too.

Clinton Vineyards dates from 1976, when its owners decided to grow the French hybrid grape Seyval Blanc. A few years later they introduced their first sparkling wine, made using the traditional Champagne method. Later, fruit-based wines were added, to general acclaim. Today the vineyard continues to feature white wines, both still and sparkling, including a fine selection of dessert wines: a delightful cassis (a black currant wine, which has won prestigious awards) and others flavored with juice from peaches, pears, apples, wild black raspberries, locally grown rhubarb and strawberries, and more. You'll certainly enjoy a tasting here, as you gaze on the landscape.

Nearby, Millbrook Vineyards and Winery is a vast, 130-acre estate, with magnificent scenery and unimpeded vistas of the distant Catskills. The vineyard, which surrounds the elegant winery, a renovated Dutch hip dairy barn, encompasses thirty acres of beautifully cultivated grapes planted in long rows that seem to stretch for miles. And you can walk through them, savoring the aroma of the grapes. With a well-run setup for visitors, including regular guided tours of both the vineyard and the winery—as well as demonstrations; tastings; a retail store with wine books, cookbooks, gourmet food products, and wine accessories; an art gallery featuring works of local artists; and even yoga classes in the vineyard—Millbrook is understandably a popular destination for wine lovers.

Some of the very best wines of the region are produced here. Among Millbrook's many accolades, it has been called "the Hudson Valley's flagship winery" by the *New York Times*, and as such, it enjoys a wide reputation. Its winemaker, John Graziano, has been here since the vineyard began its operation in 1984 and is well known in the trade. Over the years he has traveled to France to study classical French winemaking techniques in Bordeaux and Burgundy. The wines he produces—some ten thousand cases each year—include a good selection of chardonnay, Tocai Friulano, pinot noir, cabernet franc, and gamay noir. (By the way, Millbrook has another vineyard, located in California.)

After your exploration of the vineyard, be sure to take the informative tour of the winery, which will give you the complete story of the winemaking process. You can then settle in for your wine-tasting experience, following the recommendations of the experts.

45·
THE ORCHARD AND FARM
OF A HISTORIC ESTATE

Montgomery Place Restores the Original
Orchards and Gardens, Annandale-on-Hudson

ஜ DIRECTIONS

8 Davis Way, Annandale-on-Hudson, NY. Take the Taconic State Parkway to the Pine Plains/ Red Hook exit for Route 199. Go west ten miles on Route 199 through Red Hook, turn right onto Route 9G, and then turn left onto Annandale Road. Bear left again onto River Road to the entrance.

ஜ INFORMATION

Open April–October, Wednesday–Monday, 10 A.M.–5 P.M.; November, December, and March, weekends only; closed January, February, Thanksgiving, and Christmas Day. At the entrance don't miss the farm stand selling estate-grown apples and berries to help support the restoration. You can also pick your own fruit in season. Telephone: 845-758-5461. Web site: www.hudsonvalley.org/content/view/16/46/.

ஜ Montgomery Place is one of the great Hudson River estates, combining romantic, sweeping landscaped lawns, woodlands, and magnificent views. This 434-acre site includes a nineteenth-century mansion (open for visitors), a few formal gardens, magnificent trees, and a wide variety of walking trails, in addition to a working orchard and farm.

Described in an 1866 guidebook of the fine estates along this portion of the Hudson as "the most perfect in its beauty and arrangements," Montgomery Place was admired for "waterfalls, picturesque bridges, romantic glens, groves, a magnificent park, one of the most beautiful of the ornamental gardens in the country, views of the river and the mountains, unsurpassed."

Built in 1804–1805 by a branch of the Livingston family (you can visit their nearby estate, Clermont), Montgomery Place was part of a 160,000-acre family holding. The house, designed in the Federal style, was remodeled in the 1830s to reflect the elegant lifestyle of the Livingstons of the time.

It was then that the working orchard and farms and commercial nursery became part of the landscape of pleasure grounds. The splendid romantic sweep of the lawns and curving driveways and stone bridges, the plantings of groves of great trees, the variety of settings—these ideas of landscape design were made with the advice of the owners' good friend, the noted designer Andrew Jackson Downing.

In the 1930s Violetta Delafield, already an amateur botanist and expert horti-culturist (with a specialty in mushrooms), created showplace gardens at the es-tate that flourished until her death in 1949. For nearly forty years thereafter the gardens declined. Today they are being carefully restored under the direction of Historic Hudson Valley. Using Delafield's writings, oral history, old photo-graphs, and even plant orders, gardeners and landscape historians are attempt-ing to re-create her spectacularly successful gardens. (If you are interested in such detective-cum-horticultural studies, this is the place for you!)

Montgomery Place is very much a tree enthusiast's estate. There are flow-ering shrubs (lilacs in profusion in May), dogwoods, magnolias, massive horse chestnuts, maples, beeches, sycamores, and the amazing grove of giant black locust trees that surround the house.

Much of the apple orchard was planted in the eighteenth century; berries and apples grow on the same land. (It is now leased from the estate to local farmers.)

46 ·
THE AQUAPONIC SOLUTION

Cabbage Hill Farm Foundation, Mount Kisco

🎋 DIRECTIONS

115 Crow Hill Road, Mount Kisco, NY. From the Tappan Zee Bridge in Westchester, take I-287 east to exit 9A to I-684 north toward Brewster. Take exit 4 and turn left to Route 172 toward Mount Kisco. Turn right at Route 117 north (East Main Street). Continue onto West Main Street by going left on Route 133. Stay on Main Street until you come to Crow Hill Road. Turn right and continue to the farm; follow signs to the Greenhouse at Cabbage Hill Farm.

🎋 INFORMATION

Guided tours year-round, on the first and third Fridays of each month. Contact for an appointment. Telephone: 914-241-2658. Web site: www.cabbagehillfarm.org.

🎋 Considered by many people to be the answer to the world's hunger problems, aquaponics is a new and exciting development in farming. Aquaponics refers to raising fish and vegetables together without the use of earth or fertilizer. It is possible to create an aquaponic system on a rooftop in the middle of a city, as well as in a greenhouse on a farm—as at this fascinating location. (In fact, green roofs provide many advantages in addition to new sources for farming,

Cabbage Hill Farm Foundation, Mount Kisco

including insulation, and are increasingly being used by forward-thinking builders and designers.) As little as an eighth of an acre can be used for aquaponics, which is one reason it is viewed as a potential source for agriculture in crowded environments.

Aquaponics depends on a closed, recirculating system, which does not waste water (it is recycled), nor does it require additional fertilizer or any earth. The fish (often tilapia or hybrid striped bass, as here at Cabbage Hill Farm) excrete ammonia through respiration and waste, which is used to hydrate and fertilize the vegetables. Bacteria convert the ammonia to nitrite and then nitrite to nitrate, which is then a readily available form of nitrogen for plants to use.

Cabbage Hill Farm is a nonprofit foundation, devoted to sustainable agriculture. It produces many different vegetables (both year-round through aquaponics and outdoors in raised farm beds in growing season) and provides both vegetables and fish to restaurants and other venues (including its own offshoot restaurant, the inviting Pig's Ear in downtown Mount Kisco).

When you walk into the greenhouse, don't expect to find a placid, picturesque conservatory like those at nurseries or botanical gardens. This is an intense working environment. Hundreds of fish can be seen swimming around in huge open vats of constantly rushing water, while pipes overhead carry the naturally fertilized water across the center aisle to rows and rows of growing vegetables with their roots in shallow water, rather than in earth. Among the plants you'll see growing are lettuce, bok choi, tatsoi, Swiss chard, fennel, cilantro, parsley, and basil.

There are increasing numbers of aquaponic farm experiments across the country, but not all of them invite visitors. We felt fortunate to be taken around Cabbage Hill Farm by the helpful greenhouse manager, Barney Sponenberg, who will give visitors as much information on the workings of aquaponics as they may wish. While this is not a visually lovely scene, like many of the outings in this collection of farm visits, it is a fascinating one, and we recommend it to anyone with an interest in sustainable food production and new and different forms of farming.

AND IN KEEP IN MIND . . .

47 • Greig Farm, Red Hook

ADDRESS: *223 Pitcher Lane, Red Hook, NY.*
INFORMATION: Telephone: 845-758-1234. Web site: www.greigfarm.com.

This is an extensive pick-your-own farm with berries, beans, pumpkins, peaches, a winery, a greenhouse, and more. You can even pick your own asparagus for dinner!

48 • Blueberry Park, Wingdale

ADDRESS: *2747 County Route 21, Wingdale, NY.*
INFORMATION: Telephone: 845-724-5776.

Blueberry Park is devoted entirely to blueberries—forty-five acres of them! All crops are organically grown. Call ahead to find out precisely when the berries are ripe, and then pick your own. (Your kids will probably eat whatever you pick on-site!)

49 • Hilltop Hanover Farm and Environmental Center, Yorktown Heights

ADDRESS: *1271 Hanover Street, Yorktown Heights, NY.*
INFORMATION: Telephone: 914-962-2368. Web site: www.hilltophanoverfarm.org.

As of our recent visit, this was a large, thriving enterprise offering pick-your-own vegetables and fruits; educational courses and events; information on canning and pickling; and opportunities for young farmers to learn planting, composting, harvesting, and various sustainable agricultural practices. Due to unfortunate budget cuts, we do not know how many of these worthy programs will continue—hopefully all of them! This is a beautiful place for walking, with trails and vistas galore.

NEW YORK STATE
West of the Hudson

50·
AN ORCHARD IN THE SHADOW OF A STONE MOUNTAIN

Is This the West? No, It's Rockland County:
Dr. Davies Farm, Congers

✍ DIRECTIONS

306 Route 304, Congers, NY. Take the Palisades Interstate Parkway north to exit 6, to Route 303 north. Cross under the New York State Thruway and continue for eight lights to the intersection with Route 304. The entrance to the farm is on your left on Route 304.

✍ INFORMATION

Open during picking season for apples, Labor Day–November, daily, 10 A.M.–6 P.M. You can visit the farm and walk around at other times by appointment. Telephone: 845-268-7020. Web site: www.drdaviesfarm.com.

✍ This is one of the most picturesque, as well as historic, sites we have visited. For over 110 years members of the Davies family have farmed here, in the shadow of the brilliantly colored orange cliffs of Hook Mountain (which we usually see from the other side along the Hudson River, just north of Nyack). The orchard, cornfields, and tomato plantings would be exquisite in any setting, but nestled as they are, in this Wild West–looking site, they seem almost cinematic. (You expect to see a couple of cowboys riding to the top of the cliff! In fact, if the farm looks familiar to you, it might be because it has appeared in a variety of television shows and commercials, including *Saturday Night Live*.)

The orchards are on both sides of the rural highway, and you can see how old many of the trees are, with their gnarled branches and burgeoning apple crop. The owners allow visitors to walk in this enchanting atmosphere and to pick apples. There is a small farm stand by the parking area, where you can find out the details of picking or simply purchase the farm's bounty.

The original 450-acre farm stretched all the way from nearby Rockland Lake (a nice place to go after your visit) to the Hudson River. Today, there are about seventy picturesque acres (including the wonderful old orchard), an astounding size for a farm so near to the city. There are some four thousand apple trees and thirty-five acres of vegetables.

The farmhouse you see as you drive up dates from 1836; the Davies family purchased it and the original thirty-five acres in 1891. Lucy Meriwether, one of the first female doctors in the nation and the deliverer of some six thousand babies, and Arthur B. Davies, the noted painter and exhibitor at the 1913 Armory Show, married and continued their careers, as well as farming the land. Their descendants and their families have continued to farm the land, long after most of the farmland in the county has disappeared.

Visitors during apple season can go on hayrides and can picnic on-site. There is also an apple cider press. This is nature's bounty at its best!

Dr. Davies Farm, Congers

51 ·

THE FARMS AND GARDENS
OF A COMMUNAL LIFE

A Rudolf Steiner Experience at the Fellowship Community, Chestnut Ridge

℘ DIRECTIONS

241 Hungry Hollow Road, Chestnut Ridge, NY. Take the New York State Thruway (I-87) to exit 14A/Garden State Parkway; once on the parkway take the first exit and turn right onto Schoolhouse Road. At the end of the road go left onto Route 45, then right onto Hungry Hollow Road. After a quarter mile, you'll see a red barn on your left. Go up the hill to the end, park your car, and ask for directions to the office of the Fellowship Community. (It's not particularly well marked.)

℘ INFORMATION

Open to visitors all year, though we recommend May to fall, when you can pick apples and other produce. To visit the farms and gardens of the Fellowship Community you must call first to make an appointment. (At that time you will be told exactly where to meet.) You will be taken around (by a member of the community, by foot and/ or car) to see the farms, orchards, and vegetable and flower gardens. Your guide will be happy to give you information about the community and its biodynamic farming methods. Telephone: 845-356-8494, ext. 236. Web site: www.fellowshipcommunity.org.

℘ Farming and the self-sustaining community are central to many groups of like-minded individuals who choose to live together and follow a particular credo or way of life. The nineteenth-century German philosopher and founder of anthroposophy Rudolph Steiner has a devoted and worldwide following to this day; among the numerous communities committed to his teaching is this one in northern Rockland County.

You can visit this site with its many schools, farmlands, senior care center, and homes; it is a genuine community devoted to Steiner's beliefs in communal living and production of as much food as the community can manage. (Another adjacent farm was recently added to the many acres here.) Steiner's beliefs (simply put) included the rejection of materialism, the acceptance of the occult, the pursuit of "spiritual science," and a holistic vision of the universe as interconnected. His teachings, an emphasis on care for the aged, and a large system of private schools are all part of the extensive Steiner legacy. Though the members of the community do not consider themselves part of a religious order, they do espouse its spiritual values and contribute their work on the land or in the community. Here in Chestnut Ridge you can experience a farm visit that is un-like any other, because the farms and gardens are part of that community.

Although you won't be given a lecture or brainwashed in any way when you visit, it is clear that you are in an unusual place. (And there is plenty of philo-sophical material available.) A member of the community will take you around and show you the many kinds of gardens and farms, all of which are worked by members in a rather freelance kind of way. (This is not a punch-the-clock type of farm!) Members help out with planting and harvesting. (It was unclear to us whether there were specific times and chores listed somewhere.) The farms grow beautiful produce, and everyone seems to have a job to do. In fact, they have something called "Circles" (described by the community as a "unique sys-tem of spiritually integrated governance"), in which co-workers offer to be re-sponsible for a particular type of work.

There are herb and flower gardens, a kitchen garden, an apple orchard, and twenty acres of biodynamic gardens, including the Pfeiffer Center, where or-ganic beekeeping is practiced. The community includes a number of intercon-nected entities including Weleda (where natural herb remedies are made), the Waldorf Institute (where teachers are trained in the Steiner way), and schools, as well as communal eating and arts centers and even a print shop.

How close this community comes to the visions of Rudolph Steiner is not clear in this materialistic age, but certainly you will find a visit here unlike other farm outings.

The community includes about 150 acres of property, from housing and pub-lic buildings (all well integrated into the landscape) to farmland, gardens, and forest. Our guide met us at the parking area near Hilltop House, and we set out on foot to explore the grounds. We learned that about 50 percent of the veg-etables consumed by the residents of the community are grown here and that the residents themselves do much of the tending.

We first walked over to see the closest gardens of the Main Campus, an at-tractive area of buildings surrounded by colorful flower beds, lawns, and woods. Above the parking area is a large enclosure with chickens, which leads to a field planted with such items as kale, sweet potatoes, beans, peas, squash, corn, beets, chard, grapes, heirloom tomatoes, cucumbers, and mini kiwis. Near it, just above Hilltop House, where the community kitchen is located, is a heal-ing kitchen herb and flower garden. The cooks at Hilltop come here to choose whichever herbs they need for the meal being prepared. Flowers are also regu-larly picked to decorate living areas.

Our guide then drove us to a larger herb garden, where different teas are planted (mint and lovage were sighted), as well as chives and onions. Nearby are greenhouses, as well as a "kindergarten" garden, where kids learn to do their own planting.

We continued along a hilly, winding road to a more rustic area reminis-cent of farms of the past. (We were reminded that before the era of malls and developments Rockland County had many more farms of this type.) This section of the property includes the Duryea Farm, a picturesque old farm dating from the nineteenth century. Acquired by the community in the late 1990s, it contains a lovely twenty-five-acre orchard (where you can pick apples and berries), greenhouses (one that cultivates microvegetables), an animal barn, and a dairy. Here horses, sheep, and cows are very much a part of the pastoral landscape—as are the people at work cultivating the vast fields and gardens using traditional biodynamic methods (emphasizing a balance of the holistic development and interrelationship of the soil, plants, and animals as a self-nurturing system).

And where does all this genuinely fresh and organic produce go? Though, as mentioned earlier, most of it goes to serve the needs of the community dining room, some is available to members and the public at Hand and Hoe, the store outlet for the products grown on the land. (At Hand and Hoe you can also buy handcrafts and other items produced here.) This facility is open to the public every Friday afternoon, from noon to five. Hand and Hoe is located in that red barn you first see, right on Hungry Hollow Road, at the entrance to the community.

Do You Want to Learn about Beekeeping?

If you are thinking of beekeeping as a hobby or a possible career, or if you're just curious about this important subject, you can sign up for a variety of courses to see whether it's really for you. Now that many communities around the country have lifted the ban on raising bees and there is a growing desire for organic and homegrown food, beekeeping has become quite a popular undertaking. Though in recent years honeybee colonies have been mysteriously dying around the country, hives are again beginning to flourish, even on rooftops in metropolitan areas.

In the New York City region there are organizations that offer classes on beekeeping over the course of the year, but especially in the winter, in preparation for spring bees. One is Honeybee Lives, which sponsors mostly two-day weekend courses in various locations in New York State, including the Sustainable Living Resource Center in Rosendale and the Pfeiffer Center for Biodynamic Agriculture in Chestnut Ridge. There you will learn the basic requirements and responsibilities for organic beekeeping, including such topics as the naturalist approach to the nurturing of bees and the need to respect and care for them; honeybee health and disease management; how to design a proper hive for the well-being of the colony; working with swarms; and the importance of wax production. You will discover that beekeeping is a more sophisticated and complex activity than you may have realized—but, judging by the number of people who do it, it is a fascinating one, too!

For more information on courses offered and preregistration, you can e-mail Honeybee Lives at HoenybeeLives@yahoo.com.

If after your visit you wish to see and learn more about biodynamic farming, you can check out the nearby Pfeiffer Center, located at 260 Hungry Hollow Road, just down the road and across the street from the Fellowship Community. This garden of seventy vegetable and flower beds is used as an outdoor laboratory/classroom for teaching and learning about this very subject. The property is unusually pretty and also includes a small orchard (apples, pears, peaches, and quince), a greenhouse, composting facilities, and an apiary. Most of the work here is done using hand tools. The center offers internships, workshops (on such subjects as organic beekeeping and apitherapy), and various related events open to the public.

52 ·

THE OLDEST ORCHARD AROUND

The Orchards of Concklin, Pomona

🍎 **DIRECTIONS**

2 South Mountain Road, Pomona, NY. Take the George Washington Bridge to the Palisades Interstate Parkway. Continue to exit 12. Take Route 45 north. Turn right at South Mountain Road; the orchard is just ahead.

🍎 **INFORMATION**

Open year-round. You can visit whenever you like, but, of course, fruit-picking season is the best time. Telephone: 845-354-0369. Web site: www.theorchardsofconcklin.com.

🍎 This is the oldest orchard around these parts, functioning continuously since 1712. (It is also the nearest to New York City.) You'll find it is an appealing place where you can pick your own apples, as well as peaches and pears, tomatoes, peppers, zucchini, and pumpkins. (You can also buy produce at the farm market.) You are invited to walk around the very old fields and orchard, where the gnarled apple trees and wagon paths suggest the long history of this site.

53·

THE SHAWANGUNK WINE TRAIL

A French Huguenot Legacy, Orange and Ulster Counties

✏ DIRECTIONS

Benmarl Winery: *156 Highland Avenue, Marlboro, NY. Take the New York State Thruway (I-87) to exit 17 (Newburgh); then take I-84 east to exit 10, Route 9W north. Continue on Route 9W for about four and a half miles, then take a left onto Conway Road. You will see the Benmarl sign and entrance on your right, after about one mile.*

Warwick Valley Winery & Distillery: *114 Little York Road, Warwick, NY. Take the New York State Thruway (I-87) to exit 15A (Suffern-Sloatsburg) and make a left off the exit onto Route 17 north; after seven miles turn left onto Route 17A and continue for about fourteen miles into Warwick. At the intersection of Route 94 turn left. At the light take a right onto Route 1A and continue for five miles to Little York Road and turn right. The winery is one mile down the road, on the right.*

Applewood Winery: *82 Four Corners Road, Warwick, NY. Take the New York State Thruway (I-87) to exit 16 (Harriman), then take Route 17 west to exit 127 (Greycourt Road). Go south on Route 13 (also called Kings Highway; look for the grape-cluster wine-trail signs). Three miles south of the town of Sugar Loaf turn right onto Four Corners Road and go one mile. Applewood Winery is on a long dirt road on your right.*

Brotherhood Winery: *100 Brotherhood Plaza Drive, Washingtonville, NY. Take the New York State Thruway (I-87) to exit 16 (Harriman), then take Route 17 west to exit 130; take Route 208 north for about seven miles; go right on Route 94, then left on Brotherhood Plaza Drive.*

Adair Vineyards: *52 Allhusen Road, New Paltz, NY. Take the New York State Thruway (I-87) to exit 18 (New Paltz), turn left at the light, and proceed to the third light; then take a left onto Route 32 and continue for six miles; turn left onto Allhusen Road. The winery is on your right, about half a mile away.*

Whitecliff Vineyard and Winery: *331 McKinstry Road, Gardiner, NY. Take the New York State Thruway (I-87) to exit 18 (New Paltz); turn left onto Route 299, drive through the town, across the Wallkill River, and bear left onto Route 7 (Libertyville Road). Go for another two miles and turn left onto McKinstry Road. The vineyard is on the right, less than a mile away.*

Stoutridge Vineyard: *10 Ann Kaley Lane, Marlboro, NY. Take the Palisades Interstate Parkway north to Bear Mountain; at the traffic circle take Route 9W north and continue for twenty-four miles. Turn left onto County Road 14 (Western Avenue), right at White Street, first left onto Prospect Street, and second left onto Ann Kaley Road.*

✏ INFORMATION

There are eleven wineries/vineyards on the Shawangunk Wine Trail (we describe seven of these and list the others at the end of the chapter). We suggest contacting the sites that interest you and setting up an itinerary, or go with "The Little Wine Bus" (see sidebar "A Wine Country Tour").

A Wine Country Tour

If you prefer to be taken through Shawangunk Wine Country with a guide, try The Little Wine Bus, a "full-service wine tour company," as it calls itself. The company offer the tours as well as all kinds of events in the area. Tours leave from Manhattan and different parts of the Hudson Valley on weekends throughout the year. Check them out! Telephone: 917-414-7947. Web site: www.thelittlewinebus .com.

Benmarl Winery: Open January–March, daily, noon–5 P.M.; April–December, daily, noon–6 P.M.; closed on major holidays. Telephone: 845-236-4265. Web site: www .benmarl.com.

Warwick Valley Winery & Distillery: Open year-round, daily, 11 A.M.–6 P.M. Telephone: 845-258-4858. Web site: www.wvwinery.com.

Applewood Winery: Open April–December, Saturday–Sunday, 11 A.M.–5 P.M.; September–October, Friday–Sunday, 11 A.M.–5 P.M. Telephone: 845-988-9292. Web site: www .applewoodwinery.com.

Brotherhood Winery: Open April–December, daily, 11 A.M.–5 P.M.; January–March, Friday–Sunday, 11 A.M.–5 P.M. Telephone: 845-496-3661. Web site: www.brotherhood -winery.com.

Adair Vineyards: Open May–August, Friday–Sunday, 11 A.M.–6 P.M.; November–mid-December, Friday–Sunday, 11 A.M.–5 P.M. Telephone: 845-255-1377. Web site: www .adairwine.com.

Whitecliff Vineyard and Winery: Open April–December, Thursday–Monday, 11:30 A.M.–5:30 P.M.; January–February, by appointment; March, weekends and by appointment. Telephone: 845-255-4613. Web site: www.whitecliffwine.com.

Stoutridge Vineyard: Open year-round, Friday–Sunday, 11 A.M.–6 P.M. Telephone: 845-236-7620. Web site: www.stoutridge.com.

When the French Huguenots settled in New Paltz during the seventeenth century, they grew vines to make wine, as they had back in the old country. They were the first group to introduce wine on these shores, and as a result, the Hudson region has sometimes been called the birthplace of American viticulture. The wine industry grew in the region, but it was not until much later in the twentieth century that it actually flourished, as Americans developed a real taste for wine. Today there are eleven vineyards/wineries listed on the Shawangunk Wine Trail, within Orange and Putnam Counties and not far from New Paltz. A day (or weekend) trip to at least some of these makes for a delightful experience—from both the culinary and scenic perspective—especially during late summer and early fall, when the harvest takes place.

Ideally, you would explore each site by starting with the vineyard, then proceeding to the winery to see how the wine is made, and finally sampling a selection of fine vintages. Most of the wineries we recommend here have pleasant vineyards in which to walk, but a couple have instead an apple or pear orchard whose fruit is used in making a sweeter kind of wine. We propose a visit to these sites if their orchards are accessible to the public during apple- or pear-picking time.

Of these eleven vineyards, our favorite is Benmarl in Marlboro, a must-see destination. Situated atop a hill with incredible views overlooking the Hudson, this is a wonderfully rustic and natural-looking site, with thirty-seven acres of vineyard, fields, and forested picnic areas. The winery itself is housed in an unpretentious building with charming décor and picturesque views of the vineyard. The feel here is welcoming and pleasant—and refreshingly uncommercial.

We were told by the friendly winemaker who showed us around that Benmarl is also one of the oldest vineyards in the country, if not the oldest "professional" vineyard. There is a great deal of history connected with this place (you can read all about it in the brochure). Suffice it to say that the Miller family, who bought this wine-producing property in 1957, created an organization called Société des Vignerons to support further experimentation with grapevines. The results have been excellent, with the production of some of the finest wines anywhere around. Among the best are a 2005 pinot grigio, merlot, Syrah, and muscat; a 2006 chardonnay, Frontenac, zinfandel, and Riesling; and some slightly sweeter delights: Raspberry Red, sangria, and Sweet Sarah—described as rich, complex, and honeyed.

Warwick Valley Winery & Distillery offers a very different experience. First of all, this is one of the first distilleries in the Hudson Valley. Instead of a vineyard on these spacious grounds (set in a valley between two mountains), you'll find apple (not surprising, since this is apple country) and pear orchards that are open to the public for pick-your-own from early September through October. The fruit is harvested by the staff to make award-winning hard ciders, as well as fruit brandies and liqueurs. Warwick Valley is best known for these fruit-based alcoholic beverages that attempt to capture and preserve the essence of the fruit at its peak.

The site offers several extras: live music on weekends, outdoor performances on the patio, and a nice bakery and café. The mood is lively and welcoming. You'll enjoy sampling the winery's fare in the tasting room, a renovated old horse barn overlooking a picturesque pond and the orchards.

Nearby is Applewood Winery, another apple site that includes a lovely apple orchard with pick-your-own facilities for visitors. The vineyard on the property (at present not open to the public) provides some of the grapes for the winery's chardonnay, cabernet franc, merlot, and Riesling (the others are imported from across the state). We are told that the vineyard is expanding and may adopt biodynamic farming methods. For now anyway, the winery is especially known for its award-winning hard apple cider and apple wines.

This is a great place for families. While parents enjoy sipping wine, children can roam around and visit the petting zoo or take a wagon ride through the farm. The winery also offers classes in wine basics and wine tastings, along with special events such as an apple blossom festival in May and grape stomping in September.

Though our next site includes neither vineyard nor orchard, we recommend it because it is probably the most famous winery in the region. Brotherhood Winery is also described in its brochure as "America's Oldest Winery." Although this site lacks the charm and inviting walking space of neighboring vineyards, it does have the country's largest underground cellars (similar to those in famous European wineries, with their giant oak barrels all lined up in rows) and is very well set up for organized tours of its extensive facilities. Though you won't see rows of grapes growing on a sunny slope, you will see large structures where you will be able to observe the process of winemaking—which is certainly interesting and worth a visit.

Brotherhood Winery, listed on the state and national registers of historic places, has, in fact, an interesting background. In continuous operation since its founding by a European émigré in 1839, the winery was able to finesse the restrictions surrounding Prohibition by selling only sacramental wines during those stark years. Once a family business, it expanded and changed ownership in the late 1980s, with many ambitious new projects in the works. The art and technology of winemaking is very much in evidence at this well-organized operation, as are the elaborate facilities and exhibits.

The grapes used in making Brotherhood's wines come from such regions as Chile, Argentina, and the Finger Lakes, not to mention places closer to home. (In fact, a few vines have recently been planted in front of the winery.) The impressively long list of wines are considered by experts to be among the best in New York and have won more awards than those of any other winery in the state. After your tour of the premises, stop by and sample some of Brotherhood's highly regarded red and white varietal wines.

Adair Vineyards, near New Paltz, is a delightful place to visit—and quite the opposite of Brotherhood. With views of the Shawangunk Mountains, this fairly large and rustic property includes a two-hundred-year-old dairy barn (used as the winery), a ten-acre vineyard, and—to complete the picture—a stream. The French-American hybrid grapes that are used to make Adair wines come from this vineyard. (Adair also uses locally grown grapes and other fruits for its fine wines.)

This is a family venture, run by the winemaker and his wife. Their wines are handcrafted and made with great care. The ambiance is easygoing and friendly, and you can walk around the vineyard at will, without pressure. Enjoy the hayloft tasting room, where you can sample some of these artisanal varieties.

Whitecliff Vineyard and Winery is also endowed with a beautiful setting—in this case, an especially dramatic one. As you walk through the vineyard or sip wine on the deck of the tasting room, you see before you the spectacular white cliffs of the Shawangunk Ridge, which give the site its name. The vast grounds include seventy acres of farmland, which seem to stretch out forever, and a walk here is an inspiring experience.

The wines produced on the twenty-five-acre vineyard are carefully crafted varieties such as Seyval Blanc, Vignoles, chardonnay, gamay noir, cabernet franc, and pinot noir. This is a family operation, a husband-and-wife team of dedicated vintners who are committed to the quality of their operation, as well as experimenting with new methods. The results are clear: many of their wines have won prestigious awards. Be sure to sample some as you gaze on the panorama.

The Stoutridge Vineyard is distinctive for its ecological methods in wine production. It may be the only vineyard in the region—or one of few—that uses such sustainable techniques as solar panels to generate all electricity; gravity techniques (no pumps or filters) in processing the wine; and little, if any, chemical additives: no sulfites, no added sugar or water, no tannins. What you get here is as pure as wines can be.

Stoutridge is also a picturesque site, set on idyllic grounds with expansive lawns, ponds, a path through the woods along a brook, and picnic areas. Less than a mile from the Hudson, the property overlooks the river, grapevines, and fruit trees. The vineyard itself is made up of three parts, each with its own microclimate, soil, and sun exposure. As a result, there are three distinctive plantings here: a four-acre section is dedicated to Riesling, a larger section to Sangiovese (a red Tuscany grape), and a section to pinot noir and Teroldego (a northern Italian red grape).

The winery is built into a hillside on the site of a previous winery whose foundation wall remains. (It has been carefully restored and is now part of the elegant décor of the tasting room.) This hillside location has naturally kept the temperature inside at fifty-five degrees without intervention, cooling the winery in summer and warming it in winter. In addition, there is a solar-cell system facing the roof, which supplies the building with all its electrical needs.

You'll find a visit to Stoutridge not only a treat for its panoramic views and delicious wines but also interesting for its commitment to preserving a green environment.

The other vineyard/wineries included on the Shawangunk Wine Trail are

Baldwin Vineyards: 176 Hardenburgh Road, Pine Bush, NY; 845-744-2226

Brimstone Hill Vineyard: 61 Brimstone Hill Road, Pine Bush, NY; 845-744-2231

Glorie Farm Winery: 40 Mountain Road, Marlboro, NY; 845-236-3265

Palaia Vineyards & Winery: 10 Sweet Clover Road, Highland Mills, NY; 845-928-5384

Magnanini Farm Winery: Wallkill, NY; 845-895-2767

54·
A DELIGHTFUL FARM EXPERIENCE FOR ALL THE FAMILY

Lawrence Farms, Newburgh

✍ DIRECTIONS

39 Colandrea Road, Newburgh, NY. Take the New York State Thruway (I-87) to exit 17 (Newburgh); then take I-84 east to exit 10, Route 9W north. Continue on Route 9W for a mile and a half to the second traffic light and turn left onto Fostertown Road. Go one mile and turn right onto Frozen Ridge Road. Go two miles to Lawrence Farms Orchards on the right.

✍ INFORMATION

Open May, Saturday–Sunday, 10 A.M.–4 P.M.; June, Monday–Friday, 8 A.M.–6 P.M., and Saturday–Sunday, 9 A.M.–4 P.M.; July, Monday–Sunday, 9 A.M.–4 P.M.; August–late November, Monday–Sunday, 9 A.M.–4 P.M.; late November–Christmas, daily, 10 A.M.–6 P.M. Call for specific crops' availability. Telephone: 845-562-4268. Web site: www.lawrence farmsorchards.com.

✍ The twenty-four different crops grown here on these beautiful, scenic 150 acres include a panoply of picture-book examples: strawberries and peas ready in June; cherries, raspberries, and apricots in July; peaches, plums, corn, and tomatoes in August; grapes, apples, and pears in September; eggplants and

Lawrence Farms, Newburgh

pumpkins in October; Christmas trees in December—this is one of the most varied of all the farms we have visited. It is also one of the most beautiful sites we have seen in the region—stretching all the way from Orange County's hills to the Hudson River, shiningly visible in the distance.

This is a completely you-pick-it operations, and you can wander freely throughout, enjoying the picturesque, fruit-laden trees, rows of shining vegetables, and even a miniature village for the kids. There are horse-and-carriage rides on weekends, sleigh rides in winter—in fact, this is a farm designed for the public's pleasure. A visit here will both entertain the family and fill up your refrigerator with delectable produce.

55 ·

ORCHARD COUNTRY

The Joys of Fruit Picking at Applewood Orchards,
Warwick Valley Winery & Distillery, and
Ochs Orchard, Warwick

𝒫𝓈 DIRECTIONS

Applewood Orchards: *82 Four Corners Road, Warwick, NY. Take the New York State Thruway (I-87) to exit 16 (Harriman), then take Route 17 west to exit 127 (Greycourt Road). Go south on Route 13 (also called Kings Highway). Three miles south of the town of Sugar Loaf turn right at Four Corners Road and follow the signs.*

Warwick Valley Winery & Distillery: *114 Little York Road, Warwick, NY. Take the New York State Thruway (I-87) to exit 15A (Suffern-Sloatsburg) and make a left off the exit onto Route 17 north; after seven miles make a left onto Route 17A and continue about fourteen miles into Warwick. At the intersection of Route 94 turn left. At the light take a right onto Route 1A and continue for about five miles to Little York Road and follow signs.*

Ochs Orchard: *4 Ochs Lane, Warwick, NY. Take the New York State Thruway (I-87) to exit 15A (Suffern-Sloatsburg) and make a left off the exit onto Route 17 north; after seven miles make a left onto Route 17A and continue about fourteen miles into Warwick. At the intersection of Route 94 turn left. Ochs Orchard is about half a mile from the center of town.*

𝒫𝓈 INFORMATION

Applewood Orchards: Open September–October, daily, 9 A.M.–5 P.M. Winery open Friday–Sunday, 11 A.M.–5 P.M. Telephone: 845-986-1684. Web site: www.applewoodwinery.com.

Warwick Valley Winery & Distillery: Open in fall, daily, 11 A.M.–6 P.M. Telephone: 845-258-4858. Web site: www.wvwinery.com.

Ochs Orchard: Open in fall for fruit picking; June–December for vegetables. Telephone: 845-986-1591. Web site: www.ochsorchard.net.

𝒫𝓈 Spending a day picking apples (or pears or peaches) at an orchard is surely one of the pleasures of this area—so near to the urban center and yet so peaceful, so pretty, and so redolent with the aroma of nature's bounty. We loved the long rows of trees, the choosing of a perfect apple or peach for our basket, the simple long-sticked picker you are lent, the bright collection of fruit you leave with—all part of the adventure.

Visit any of the orchards we describe here for unique experiences; each has a slightly different emphasis: one is more devoted to cider and fruit wine, another to family picking, and so on. But all of them provide a quintessential autumn outing.

Warwick, by the way, is particularly noted for its fertility and the success of its crops—especially apples. Read the following descriptions for the special flavor of each orchard, and choose your outing accordingly.

Applewood Orchards

On a crisp autumn day one of our favorite walks is through the orderly rows of apple-laden trees in Applewood Orchards and Winery. These 160 acres of rolling orchard hillside with distant valley views are an inviting spot for strolling at will, picking apples (several kinds), and admiring the Grandma Moses–style scene. This is also a winery, with tasting and other events. (For information visit www.applewoodwinery.com.) A small pond, picnic table, and brilliant red apples weighing down the trees give this setting an old-fashioned American look and feel.

This is the perfect family outing, not only because children and adults find apple picking fun but also because the gentle slopes, dozens of grassy lanes between the trees, and fresh clean air make walking very easy. There are tractor and wagon rides, too, for those who don't want to stroll.

When you arrive at Applewood, you'll see a small barn where you can pick up empty apple bags and a map of the orchard, as well as farm produce. The map lays out the area of the orchard, including where to find trees of McIntosh, Cortland, Rome Beauty, Red Delicious, and Golden Delicious apples. You can either plot your route according to the kinds of apples you like or head down the driveway to the oldest house in Orange County, the Staats House, which was built in 1700 and is at one end of the orchard, and start from there. Stopping for a look over the valley below, you can begin your stroll in the Rome Beauty section of the orchard and proceed along, plucking apples off the trees as you wish. (Apple pickers—long-handled baskets—are for rent if you're serious about reaching the high ones.) Either walk straight through the orchard or zig and zag as you wish, ending up at the pond that we found nestling below the trees and surrounded with children who had just come from a tractor ride through the orchard. The worker in the barn will estimate the weight of your bag of apples and charge you (very reasonably). If you have picked no apples, that's all right, too.

In addition, Applewood makes wine: it has vineyards and produces "limited edition" wines, including oak-aged reds and barrel-fermented chardonnays.

Warwick Valley Winery & Distillery

Here you can not only pick your own apples and pears but also sample the many fruit wines, hard apple cider, and chardonnay and Riesling wines that the

Warwick Valley Winery & Distillery, Warwick

winery produces. (There are a number of organized tasting events here: you can try the pear brandy or the raspberry cider, for example.)

The setting is attractive and bustling, and you'll enjoy the combination of fruit orchard and winery. Warwick recommends that you call in September for pear week and particular apple information.

Ochs Orchard

Ochs Orchard offers a greater range of tree fruits. Here you can bring home a basket of self-picked delicacies, ranging from plums to pears, nectarines, apricots, and even two kinds of grapes: Concord and Niagara. There are also traditional American crops such as Indian corn, gourds, cucumbers, peppers, carrots, and tomatoes. The orchard produces its own apple cider and has a farm stand selling fruits and vegetables for most of the year.

It is fun to combine the many kinds of tree fruits (call to find out when each is ripe, and plan your trip according to your taste) and to see how each grows in its own area of the orchard.

Cristina's Apple Surprise

Serves 6
1 cup raisins
3 tablespoons red wine
6 large apples
rind of one lemon, grated
juice of one lemon
6 tablespoons honey
6 tablespoons apricot preserves or red currant jelly

In a bowl, soak the raisins in the red wine for one hour. Preheat the oven to 375 degrees. Wash and core the apples, but don't peel them. Cut off a thin strip at the base so the apples will remain upright. Put them in a baking dish and fill the center with the wine-soaked raisins and the grated lemon rind, lemon juice, and honey. You can prepare this much ahead of time, cover with plastic wrap or aluminum foil, and set aside until ready to cook.

Pour one-half cup of warm water around the apples, cover with a sheet of aluminum foil, and bake for one hour in a 375-degree oven. Remove the foil, pierce the skin of each apple two or three times with a sharp knife, and bake for ten to fifteen minutes longer. Transfer the apples to a serving dish. Stir the preserves or jelly into the cooking-pan juices and pour over the apples. Serve warm or cold.

This is the perfect dessert to serve (especially) in fall, during apple season.

56·

CHERRY PICKING, NECTARINES, AND OTHER PLEASURES

Prospect Hill Orchards, Milton

❧ DIRECTIONS

Milton, NY. Take the New York State Thruway (I-87) to exit 17 (Newburgh); then take I-84 east to exit 10, Route 9W north. Continue on Route 9W for ten miles to Milton Turnpike. Turn left at the light and follow the yellow signs.

❧ INFORMATION

This farm has several locations depending on what you plan to pick. Cherries are picked at 340 Milton Turnpike, apples and pears at 40 Clarkes Lane, and peaches at 125 Milton Crossroad. Cherries ripen in mid-June; August brings tree-ripened peaches and nectarines; September and October are the time for apple picking. Telephone: 845-795-2383. Web site: www.prospecthillorchards.com.

❧ If you've never been to a cherry orchard—or a peach or nectarine orchard —the time is ripe! You can enjoy this one starting in mid-June, when the first cherries are ready. A wagon ride will take you to the orchard, and after you've picked your own bounty to take home, "you can escape the summer sun while picnicking in the shade of the cherry branches," according to the owners.

There are picturesque orchards of peaches and nectarines—August is the time for them—and apples in fall, of course. By September the trees are laden with over a dozen different varieties of apples—ranging form Macoun to Golden Delicious, McIntosh to Fuji, and many, many others—their deep reds and yellows decorating the low-hanging branches.

Prospect Hill Orchards has been in business and in the same family since 1817 —seven generations! (Despite its antiquity, we should add that today's owners use modern, low-spray techniques.) The orchard offers all kinds of welcoming events, from tours to farmers' markets to hayrides. There is even a pumpkin totem pole. Obviously the good folks at Prospect Hill know what families with children will enjoy when making a farm visit, but in our eyes, nothing equals the simple beauties of fruit-laden trees in all their glory.

57·
THE HYDROPONIC WAY

Abundance Growing in Water at Mountain Fresh Farms, Highland

❧ DIRECTIONS

282 Orchard Road, Highland, NY. The owners will give you directions when you telephone.

❧ INFORMATION

Open year-round. This farm can only be visited by appointment (but the owners are most welcoming). Telephone: 845-795-2260. Web site: www.mountainfreshfarms.com.

❧ This is truly an unusual experience, and we recommend it to anyone interested in the growing process. When you visit, you will be guided through the greenhouses, where everything is grown—without earth! You will see for yourself the hydroponic system, which is of ancient origin (dating to 600 B.C.), and you will without a doubt admire the giant red tomatoes, the brilliant, huge orange peppers, and the vast rows of fragrant basil in different stages of growth. In this computer-controlled environment, vegetables and herbs are grown all year long in a carefully controlled habitat that feels and smells exceptionally clean and pure. (There are no bugs, and the climate is always ideal.)

The term *hydroponic* means growing plants without earth. Instead of soil, natural products called perlite and rockwool are used; both retain water in vast irrigated trays or channels. No pesticides are used (or needed), and natural sunshine permeates the greenhouses. The water is constantly tested. (You will be pleasantly surprised to discover that your own tristate supermarket may carry Mountain Fresh Farm's produce, including the oversize tomatoes and the thriving basil in small cellophane, triangular-shaped bags).

This is a unique place to visit; the owners are enthusiastic and knowledgeable and will answer any and all questions. But you cannot purchase their produce. Although they grow cilantro, cucumbers, mint, oregano, and sage, in addition to the other produce mentioned, this is a wholesale farm, so don't expect to take home their vegetables—even if you'd like to. However, you'll go home quite amazed by this rare experience, since hydroponic farms are few and far between.

Mountain Fresh Farms, Highland

58 ·

IN A TRADITIONAL
NEW ENGLAND STYLE

DuBois Farms Offers a True Farm Experience,
Highland

❧ DIRECTIONS

209 Perkinsville Road, Highland, NY. Take the New York State Thruway (I-87) to exit 18 (New Paltz) and take Route 299 for five and a half miles. Turn right onto Route 9W and go for about four and half miles, then take a right onto Perkinsville Road. Go for less than a mile to the farm.

❧ INFORMATION

Open end of August–mid-November, daily, 10 A.M.–5 P.M. Call for information on crop schedule and festivals. Telephone: 845-795-4037. Web site: www.duboisfarms.com.

❧ When you first come to Dubois Farms, pick up a map, which will point you to the many and various growing areas on this spacious and charming farm. Depending on when you visit, you can picnic, see antique farm equipment,

Apple blossoms (iStockphoto.com/Dwight Nadig)

enjoy a hayride, get lost in a corn maze, or pick your favorites from among twenty-three types of apples, three kinds of pears, and grapes—which you can turn into your own delicious vintage.

This farm truly celebrates nature's bounty in late summer and early fall, with such events as Capital Plum-A-Palooza, Grapes Galore Festival, Scarecrow Jamboree, and Nectarine Mania. If celebrations or fruit picking are not your thing, you will nevertheless love the atmosphere of lush produce and nature's offerings.

59·
LAKESIDE PICKING

Peaches and Apples at Weed Orchards, a Family Farm, Marlboro

❧ DIRECTIONS

43 Mount Zion Road, Marlboro, NY. Take the New York State Thruway (I-87) to exit 17 (Newburgh); then take I-84 east to exit 10, Route 9W north. Continue on Route 9W for seven miles to Western Avenue (across from Sunoco). Take a left and continue two miles to four-way stop. Take a right onto Lattintown Road. Go two miles to Mount Zion Road. Take a left and continue a quarter mile to the orchard.

❧ INFORMATION

Open August 15–November, daily, 11 A.M.–5 P.M. Telephone: 845-236-2684. Web site: www.weedorchards.com.

❧ This vast, hilltop orchard has its own sparkling lake and spectacular views of the Marlboro Mountains. In season you'll find at least fifteen varieties of apples, as well as mouth-watering peaches, which you can pick yourself. Pears, grapes, pumpkins, beans, peppers, squash, and eggplants all grow here. Many family activities are offered, such as hayrides, a petting zoo, and other wholesome pleasures that the kids will love. This is a real farm—rustic, unphony, and inviting.

60·
A SPECIAL TREAT

*Picking Cherries and Plums at Dolan Orchard in
Ulster County, Wallkill*

❧ DIRECTIONS

1166 Route 208, Wallkill, NY. Take the New York State Thruway (I-87) to exit 17 (Newburgh); then take I-84 west to exit 5, Route 208, which you take north to Wallkill. The farm is on your left as you go north.

❧ INFORMATION

Open daily, 9 A.M.–5 P.M.; for picking cherries, early June–early July; for plums, July–August. Call before you go for specific information. Telephone: 845-895-2153.

❧ In the shadow of the Shawangunk Mountains, you'll find this rustic, picturesque orchard, where you can pick (or buy) delicious fruits. In addition to the plums and cherries, there are also apples, pears, and pumpkins grown here. Although there is a farm stand directly along the highway, the orchards themselves are well off the road and are an ideal place for family picking in a pleasant rural setting.

61 ·
FROM HISTORIC FAMILY FARM TO COMMUNITY GARDENS

The Rebirth of Cropsey Farm, New City

✿ DIRECTIONS

220 South Little Tor Road, New City, NY. Take the Palisades Interstate Parkway to exit 10; turn left onto Middletown Road, then right onto South Little Tor Road. Continue for several blocks. The farm is on your right, directly across from Van Houten Nursery and Gardens.

✿ INFORMATION

Rockland Farm Alliance: Telephone: 845-634-3167. E-mail: info@rocklandfarm.org. Web site: www.rocklandfarm.org.

✿ Since the 1970s, when farms in this region began to be sold at an alarming rate to housing developers, there have been fewer and fewer of them. For example, in the 1920s Rockland County was a flourishing agricultural community with some nine hundred farms; unfortunately by 2003 only a handful of them remained. In recent years a new trend has emerged—and not only in Rockland County, where the Rockland Farm Alliance (including farmers, community activists, county officials, and local citizens) has been created to encourage local sustainable agriculture. Throughout the tristate region and well beyond, a local food movement in farmers' markets and other direct sales to consumers and restaurants has appeared. More and more young people are looking to pursue farming as a viable career, and some long-abandoned farms are even being rehabilitated.

In Rockland County, Cropsey Farm is an example of this new revitalization of farming and its sustainability. Five acres of this once-large (fifty acres) farm

---

What Is a CSA?

CSA stands for "community supported agriculture." With more and more people wanting to grow their own produce (without owning enough land) cooperative farming is thriving. The would-be farmer buys shares in a CSA farm, and in return a certain amount work is required. Most CSA farms in this area use organic methods to produce sustainable—and always the freshest—vegetables and fruits. (We even discovered an orchard that lets people lease one of its apple trees!)

maintained by the same family since the eighteenth century are being transformed into a community farm (a CSA). As of this writing, the operation is still
in its initial stages, but it is hoped that Cropsey Community Farm (as it is now
called) will be in full operation by 2012. This is to be the first community farm
in Rockland County. The plan is to include both a working, money-making organic farm and an education/demonstration area. This has been an ambitious
project, involving community and local governments, as well as many energetic
and forward-thinking individuals committed to nature and the environment
(Judging by their dedication and willingness to be inventive, we are confident
that the project will come to fruition).

The property itself appears like an oasis of peace and serenity along a busy
suburban street. Including fields of rich farmland, fine old trees, red barns, and a
beautiful 1769 native-red-sandstone farmhouse still occupied by members of the
Cropsey family, it is as bucolic a spot as you'll see in this area. (Directly across
the street you'll find a prosperous-looking garden center/nursery, once part of
the farm.) When we visited, we saw a group of young people—staff members
and volunteers—assisting the recently hired farmer in planting rows of garlic,
apparently their first efforts in cultivating the land after its many years of in-
activity. It was interesting to see this "work in progress" and to imagine what
it would look like when finished (though anything to do with nature is never
really "finished"). There was great enthusiasm in the air, as the project had just
been approved by the various authorities, and everyone involved seemed eager
to get to work planting, finding new volunteers, raising money, bringing high-
quality organic produce to the community, and establishing an ambitious assortment of programs. We hope that you will enjoy visiting this wonderful and
sure-to-be-bountiful site!

AND KEEP IN MIND . . .

62 ◆ Lyonsville Sugarhouse, Accord

ADDRESS: *591 County Route 2 (Krumville-Kripplebush Road), Accord, NY.*
INFORMATION: Telephone: 845-687-2518. E-mail: lyonsvillefarm@yahoo.com.

You can see how maple sugar and maple syrup are made by taking a tour here between mid-February and April. Be sure to call first.

63 ◆ Jenkins-Lueken Orchards, New Paltz

ADDRESS: *69 Yankee Folly Road (off Route 299), New Paltz, NY.*
INFORMATION: Telephone: 845-255-0999. Web site: www.jlorchards.com.

This large farm/orchard is open all year round. (But we recommend going during apple season, when you can pick your own, including our favorite, Macouns.) Honey and cider are also produced here, as are pumpkins, pears, peaches, and many vegetables.

Orchard at harvest time (iStockphoto.com/Denis Sauvageau)

NEW JERSEY

64 ·
TWO THOUSAND HERBS!

A Visit to Well-Sweep Herb Farm, Port Murray

✿ DIRECTIONS

205 Mount Bethel Road, Port Murray, NJ. Take I-80 to exit 19. At the end of the ramp, turn left onto Route 517 south. After about five miles take Grand Avenue (which intersects with Route 46/ Main Street); continue for another mile until Grand Avenue becomes 629 south/Rockport Road, then go another four miles. After a steep climb turn right onto Mount Bethel Road. You'll find the farm one mile ahead, on the left.

✿ INFORMATION

Open Monday–Saturday, 9 A.M.–5 P.M. (closed Sundays, except in December, when it is open 11 A.M.–4 P.M.). Closed on major holidays. Call before going during the winter months, January–March. The farm offers tours of the grounds and greenhouse, such as winter greenhouse tours and medicinal herb walks; workshops and how-tos on such topics as organic gardening and dried flower arrangements; and open houses and guest speakers. Call for availability and hours of these special offerings. Telephone: 908-852-5390. Web site: www.wellsweep.com.

✿ What to look for: an amazing assortment of every herb imaginable, both indoors and outdoors; many display gardens (including a formal herb garden and knot garden); and a gift shop (great for browsing!) filled with dried flower arrangements, wreaths, potpourri, craft items, and garden/botanical books.

Any herb enthusiast—or, for that matter, anyone at all—will be captivated by Well-Sweep Herb Farm, an enchanting place to visit. Picturesquely set on 120 rural acres in the rolling hills of northwestern New Jersey, it features an impressive herb collection, close to two thousand varieties at last count, as well as a fine assortment of perennials. With the notable exception of the quarter-acre formal display garden (in full view by the side of the road), most plantings are informally arranged, in keeping with the relaxed mood of the place and its staff. (In fact, Well-Sweep seems much more low-key than most other commercial nurseries.) You are free to bring your picnic and wander about the spacious grounds at your leisure, discovering (besides family dogs and chickens roaming about) the farm's wide range of offerings, especially the plentiful varieties of herbs, both common and rare, such as silver corkscrew chives, blackberry lily, sweet woodruff, fragrant heliotrope, sweet cicely, milk thistle, lemon spice, and lippia dulcis—not to mention an abundance of lavender, one of our favorites.

The story of Well-Sweep Herb Farm began in 1966, when Cyrus and Louise Hyde, both passionate about gardening and farming, first discovered this

Well-Sweep Herb Farm, Port Murray (Courtesy Well-Sweep Herb Farm)

property. Over these many years they have continued to pursue their interest in developing plant varieties; the result is this remarkable place, where the focus seems to be as much on inviting people to learn more about plants and their role in our world as on any commercial venture.

Both Hydes are still very much involved in the actual running of the farm, as you will undoubtedly notice. Louise is often busy walking about, checking on the farm's many ongoing operations, while Cy can usually be found working in his garden. Their complete and continuing dedication to plants is obvious—and infectious.

On your visit you might like to hear what Cy has to say on the subject of herbs, about which he is a recognized authority. (In 1991 the Herb Society of America gave him a special award for excellence in horticulture.) He is happy to share his information as he takes people around the farm on informal walks, peppering his talks with tales of the land. He explains that "herbs come under four categories: medicinal, culinary, dyeing, and fragrance," something we may have known but not thought about lately. Not surprisingly, he is a fan of herbal remedies, adding that "people are trying to get away from a lot of the different medicines that have side effects; they want to go back to nature." He talks of the use of herbs dating to ancient times, saying, "We can tell the ancient Egyptians used herbs because we've identified certain plants in their paintings and drawings."

You can also learn about growing your own herbs, as you either take a walking tour or attend a class, lecture, open house, or herb festival, all of which are offered on a regular basis (call ahead for information). There are tours of culinary, medicinal, and formal herb gardens and minicourses on how to make your own herbal condiments or how to harvest plants for teas (as well as sampling various herb tea blends), how to make homemade soap (from herbs, of course!), how to make lavender wreaths, and much more.

Other things to enjoy while at Well-Sweep include the butterfly, medicinal, perennial, and rock gardens (some of which have appeared in national magazines) and the brick-pathed herb garden, with its charming knot garden and displays of 36 basil plants, 54 rosemary plants, 108 thyme plants, and more! And don't forget to visit the greenhouses, where you can find expert plant propagators doing their thing—another interesting activity to observe. Most of all, enjoy the pleasure of just being in this spot: its gentle fragrances and soft colors and its serene environment, far from the congestion of urban life.

65.

TAKE YOUR PICK OF RASPBERRIES, STRAWBERRIES, OR PUMPKINS

An Invitation from Sussex County Strawberry Farm, Newton

❦ DIRECTIONS

565 Route 206, Newton, NJ. Take I-80 west to exit 26 (Route 206) and drive north for about nine miles. The farm is on your right, between Andover and Newton.

❦ INFORMATION

Open in season, weekdays, 7 A.M.–8 P.M.; Saturday, 7 A.M.–5 P.M.; Sunday, 7 A.M.–4 P.M. It's best to call ahead for an update on picking conditions. Telephone: 973-579-5055. Web site: www.sussexcountystrawberryfarm.com.

❦ The Sussex County Strawberry Farm is a pleasantly rustic spot featuring long rows of low-to-the-ground strawberry and raspberry bushes. The views are quite vast—there are few trees on the farm—and the ambiance is inviting and relaxed. This is an ideal place for families with children in tow; besides the fruit picking (which in itself can be entertaining for all concerned), there are hayrides, pony rides, and various other child-oriented amusements, especially in the fall.

The owners have been here for more than thirty years, and their experience in operating such a farm is clear (though relaxed, the operation is well run and works efficiently). They list tips on how best to pick fruit and what to look for, which is in itself a worthwhile learning experience. Here are some of their suggestions:

- It's best to pick berries (or any fruit for that matter) early in the day or late afternoon, when it's cooler.
- Pick only ripe berries, as green ones will not ripen once picked. One way to tell whether raspberries are ripe is by how easily they separate from the plant. If they don't separate easily, they are not ready to be picked.
- In picking raspberries (which are especially delicate), use your thumb, index finger, and middle finger; the berries must be carefully placed directly in a basket (or other container), never dropped. When picking strawberries, look under the leaves of the plant to find the berries; carefully pinch the stem to make sure that the hulls remain on the berries.
- Once picked, all berries should be put in a shady spot or a cool storage area, so they can be kept fresh for a couple of days.

66·
A KOREAN ORCHARD OF UNUSUAL BEAUTY

The Magical Sight of Evergreen Farm,
Hamilton Township

❧ DIRECTIONS

1023 Route 524 (Yardville-Allentown Road), Hamilton Township, NJ. Take the New Jersey Turn-
pike to exit 7A and go east on I-195 to exit 8. Take Route 524 west to Allentown; the farm is just
past Allentown.

❧ INFORMATION

Open daily except Saturday. Call for the best time to see Korean peaches, pears, plums, and grapes growing. Telephone: 609-259-0029. Web site: www.evergreenfarm.us.

❧ We have visited many an orchard in compiling this book, but none so lovely as this one. The Korean owners grow fruit in a different way from the usual American lines of trees; here the trees are grown atop rows of identical metal arches in great tunnels that are delicious smelling and visually stunning. The pear trees, for example, are trained to stretch across these arches; the owners have worked for many years since starting their farm to create this exquisite

Evergreen Farm, Hamilton Township

Evergreen Farm, Hamilton Township

—and apparently very successful—farm/orchard. These arches of fruit trees resemble a great Gothic church when you walk under them, but there are bright, hanging fruits everywhere, with a low bed of grass beneath your feet. You will be astonished that you are allowed to pick the fruit yourself from what almost seems too beautiful to touch.

About 70 percent of the orchard is devoted to Korean pears (ripe in early to mid-September), but there are also plums and peaches growing, as well as jujube fruit and Chinese cabbage. The pears are of particular note; they are different from American pears, with a sweeter, juicier taste. There is a big specialty market in New Jersey nowadays for such fine ethnic foods, and Evergreen Farm apparently supplies a goodly amount of the fruits.

We thoroughly enjoyed our time at this orchard, not only because it's a beautiful place but also because we happened upon dozens of workers eating together at an outdoor table, being served Korean delicacies (including, no doubt, the picture-perfect fruit growing just behind them!).

67·

AN AUTUMN FAMILY DAY
ON THE FARM

An Inviting Apple Orchard, a Cider Mill,
and a Corn Maze at Hacklebarney Farm,
and Bountiful Picking, Another Maze, and
a Hay Tunnel at Alstede Farms, Chester

⚘ DIRECTIONS

Hacklebarney Farm Cider Mill: *104 State Park Road, Chester, NJ. Take I-80 west to exit 27A. Go south on Route 206 to Chester. Turn right on Route 513 (Main Street) and then left on State Park Road. Look for the farm on your right.*

Alstede Farms: *80 Route 513 South, Chester, NJ. Take I-80 west to exit 27A. Go south on Route 206 to Chester. Turn right on Route 513 (Main Street) and go three-quarters of a mile to the entrance.*

⚘ INFORMATION

Hacklebarney Farm Cider Mill: Open in fall, during apple season, early September–November, daily except Monday, 10 A.M.–5 P.M.; end of November–end of December, daily, 10 A.M.–noon. Telephone: 908-879-6593. Web site: www.njcidermill.com.

Alstede Farms: Open year-round, Monday–Saturday, 9 A.M.–6 P.M.; Sunday, 10 A.M.–5 P.M. Fall hours: 9 A.M.–8 P.M. daily. The best time to visit is in late summer/early fall. Telephone: 908-879-7189. Web site: www.alstedefarms.com.

⚘ For a perfect fall day's outing we take you to the picturesque Hacklebarney Farm Cider Mill, in Morris County's alluring Black River valley. If you particularly like apples and cider, that's all the more reason to visit! Beginning around Labor Day (when the apples begin to ripen) through the end of fall you are welcome to enjoy a leisurely stroll through the farm's hillside orchard, to stop at the cider mill itself to witness the intriguing process of making cider, and no doubt to sample some, too. And to complete your autumn day away, you can also pick pumpkins in the nearby fields and experience firsthand the farm's wonderful corn maze, which gets bigger and more ambitious each year.

The farm is situated along a hilly and windy road, in a region first settled by English colonists in the mid-1700s. The area of Chester (named because many of these folks came from the English town of Chester) became an important way station, where travelers stopped en route to the Delaware Gap from the eastern parts of the state. Its commercial enterprises included apple farms, where cider was made and distilled into applejack and apple brandy (reputed to be unusually tasty!). Hacklebarney Farm actually dates to the mid-nineteenth century, when one of the direct descendants of the early settlers purchased about 150 acres of farmland here.

Begin your visit by walking through the lovely orchard, which has more than 350 acres, with some trees in it well over a hundred years old. You'll find an abundance of different varieties. (We're told there are at least twenty-three.) The picture-book red barn just below, converted to a cider mill, was built in the 1850s. Here, the farm's famous wood-pressed apple cider is made from an exclusive blend of homegrown apples. Demonstrations of the cider-making process in action are offered to the public usually on weekends at noon (but call first to make sure). The cider is made using an oak rack and cloth process, pressing slowly to ensure a pure cider without sediment. We recommend it!

You are free to walk around further, to check out the four-acre maze (particularly if you have children along), to find a few pumpkins to pick for your Halloween festivities, and to leave with delicious items from the shop to savor on your way home!

Alstede Farms is one of the busiest and most family-oriented pick-your-own farms in New Jersey. This vast (360-acre) property, located in a picturesque hillside setting, includes acres of fruit trees and bushes, planted in neat rows, as well as inviting displays of vegetables and flowers—all there for your picking pleasure! The farm is a happy, sunny place, always buzzing with activity.

When you first arrive, you are greeted with such a choice of different options of things to see and do that you almost don't know where to begin. In addition to the delicious-looking fruits and vegetables (especially profuse in late summer and early fall), there are several nature-related amusements that are especially geared to children. A ten-acre corn maze stands out; when we visited, its theme

Farm Mazes

Mazes are versions of the topiary, the ancient art of shaping plants into living sculptures. Dating back at least to the Middle Ages, they were a popular feature in the elegant garden designs of the Renaissance. But the rustic corn maze is a modern concept. And despite the fact that such mazes last a relatively short time —only during the couple of months of fall—and that they require intricate planning and maintenance, they seem to be gaining visibility around the country.

Mazes are being designed in all sorts of fanciful ways these days. Some feature realistic pictures: a farm scene or an object such as a barn. Others feature geometric abstractions and problematic designs with twists and turns and dead ends —easy to enter but a challenge to exit.

Alstede Farms, Chester

was ice hockey, in honor of the New Jersey Devils. Also in prominent displays are farm animals for petting and a giant hay-bale pyramid with tunnels—always popular with kids. On fall weekends there are pony rides and farm tours on a hay wagon.

Pick-your-own produce includes strawberries from Memorial Day to the Fourth of July; black and purple raspberries and green and pink gooseberries from the end of June to late July; currants and blueberries from August to mid-September; peppers, tomatoes, and eggplants from late July to October; apples after Labor Day through October; pumpkins and gourds in September to mid-November; and flowers (especially zinnias and marigolds) from mid-July to October.

In addition, Alstede Farms offers many events. Every year the farm turns ten acres of cornfield into a maze with a new theme.

- **Fall Family Fun Days**, weekends from early September to early November: activities for children include scenic hay-wagon rides, tractor-train rides, a moonwalk, plus the giant hay-bale pyramid, bale tunnel, and farm animals —everything included. You can also enjoy picking apples and pumpkins, all for the same price.
- **A Harvest Moon Hayride**, weekends of mid- to late October: again, this is a family activity, including a nighttime corn maze (bring along a flashlight), harvest moon hayride, country music, campfire, and unlimited hot and cold cider to drink.

Alstede is obviously a well run and organized farm. When you visit, first stop by the main building (which houses the indoor/outdoor market) to pick up information and a map. The friendly staff will inform you as to which produce is available to pick at the time, provide you with what you need, and direct you to the appropriate locations. A newsletter with the latest farm information is also given out. You'll find that a day here (with or without kids) is a very pleasant —and delicious!—experience.

68 ·

NOT FAR FROM THE CITY

*Peaches Galore and Hayrides, Too, at
Demarest Farms, Hillsdale*

✍ DIRECTIONS

244 Wierimus Road, Hillsdale, NJ. From the George Washington Bridge take Route 4 to the Garden State Parkway north. Take exit 168 to Route 502 (Washington Avenue); make a left and go to the intersection with Wierimus Road. Turn right and continue to the corner of Hillsdale Avenue. Demarest Farms headquarters are on your right (although the orchard is not visible from the street). It is well marked, and you can't miss it.

✍ INFORMATION

Picking peaches and apples takes place Friday–Sunday, noon–4 P.M. Call first for what's in season and for farm events. Telephone: 201-666-0472. Web site: www.demarest farms.com.

✍ You'll find Demarest Farms a bustling, cheerful, and friendly place that offers a variety of activities and produce: apples, peaches, and pumpkins (in a beautiful orchard across the street from the farm market, away from the crowds); a wide selection of fruits, vegetables, and herbs, as well as flowers, hanging baskets, garden items, and plants to purchase; and, if you're hungry, a deli, country bakery, ice cream parlor, made-from-scratch salad bar, jams, jellies, cheeses, and delicious homemade doughnuts and cider. In season there are also hayrides, in addition to a variety of special events (call for dates).

But what makes Demarest Farms special is that it's among the oldest and most venerable farms in the region. The Demarest family, which traces its roots to seventeenth-century France, settled in Bergen County centuries ago; part of this land was purchased in 1886, and many generations later the family still owns and runs it. In fact, during your visit you are likely to see and perhaps chat with the present owners, who are actively involved in the daily workings of the farm.

Originally Demarest Farms was a traditional forty-acre farm with the typical farm animals. It wasn't until 1916 that the fruit orchard was begun, soon followed by a small roadside fruit stand, "Annie's Fruit Stand in Hillsdale, NJ," as it was known by its many enthusiastic patrons. Over time the variety of fruits and vegetables grown and sold here has multiplied to include about every kind you can think of. And in the 1970s the pick-your-own concept was introduced, to great popular acclaim.

We especially enjoyed walking through the orchard—a real treat, whether you actually pick the fruit (about twenty kinds of apples) or simply take a stroll amid the trees laden with luscious and fragrant apples (in fall) and peaches (in summer). The apple orchard alone includes some seventeen-plus acres, and the adjacent peach orchard, about eleven and a half acres, so this can be a nice, long, and picturesque walk.

Demarest Farms, Hillsdale

69·

A NEW JERSEY SPECIALTY

Exploring the Cranberry Bogs at
Double Trouble State Park, Lacey Township;
and Brendan T. Byrne State Forest, New Lisbon

℘ DIRECTIONS

Double Trouble State Park: *Double Trouble Road and Pinewald Keswick Road, Lacey Township, NJ. Take the Garden State Parkway south to exit 80. Turn left (south) onto Double Trouble Road and go to the end of the road (about three miles). The park entrance is located directly in front of you, at the intersection with Pinewald Keswick Road. Cranberry Village is just beyond the entrance.*

Brendan T. Byrne State Forest *(main entrance and visitor center): Route 72, New Lisbon, NJ. Take the Garden State Parkway south to exit 67, take Route 554 west, and continue onto Route 72 west. The entrance is one mile east of the intersection with Route 70, on your right. Stop at the visitor center for maps and other information.*

℘ INFORMATION

Double Trouble State Park: Open daily from dawn to dusk. Admission is free. Telephone: 732-341-6662. Web site: www.state.nj.us/dep/parksandforests/parks/double.html.

Brendan T. Byrne State Forest: Telephone: 609-726-1191. Web site: www.state.nj.us/dep/parksandforests/parks/byrne.html.

℘ The Pine Barrens is one of the greatest expanses of wilderness in the Northeast, including more than one million acres of dense forests, swamps, rivers, and pitch pine lowlands. New Jersey was—and is—an important cranberry-producing state (it's third after Massachusetts and Wisconsin), a statistic that may not be widely known. And much of this activity has taken place on these very sites. So if you like walks combining history with nature, head for Double Trouble State Park and Brendan T. Byrne State Forest, where you can experience the beauty of the Pine Barrens (without committing to an arduous hike), along with a walk through historic villages.

The curious name "Double Trouble," which began with the cranberry industry, may have one of two origins: a certain Thomas Potter said those words in disgust after heavy rains washed out the dam twice; or the words were shouted out by workers after discovering two leaks in the dam, presumably caused by gnawing beavers.

Garnet's Easy Cranberry Crunch Recipe

Serves 8
2 cups fresh cranberries
¼ cup sugar
¼ cup chopped pecans or walnuts

For topping:
1 egg
½ cup sugar
½ cup flour
1 stick of melted butter

Preheat the oven to 325 degrees. Rinse, drain, and dry cranberries, and spread on a pie plate; sprinkle them with nuts and sugar.

For topping: In a bowl, beat the egg well and add sugar, mixing thoroughly. Add the flour and melted butter, and mix well. Pour the mixture on top of the berries.

Bake in the oven for about forty-five minutes. Serve with vanilla ice cream or whipped cream.

Cranberry farming was particularly suited to this area of cedar forests and flowing streams; as trees were cut down, cleared swamplands were planted with cranberry vines. By the early part of the twentieth century, the cranberry industry was thriving here under the Double Trouble Cranberry Company, and the village was in full operation. Until the 1960s, when farming demands and methods began to change, some of the state's largest harvests came from these bogs. Today only a few of the remaining bogs are harvested, under special lease arrangements with local growers.

The Double Trouble Historic District, as the abandoned village and its immediate surroundings are known, is situated on high ground between two bogs (one is now dry) and provides you with a view of what a cranberry-production community was like. Before exploring the village (it makes more sense to do that first and then to take your nature walk), stop at the small office on the right and pick up a map that identifies each building (thirteen are listed), as well as a detailed nature trail guide. The village is very rustic and the impression refreshingly uncommercial—as of now, anyway, there is no entrance fee and nothing at all to buy. You can just wander about at your own pace, enjoying this small settlement as it more of less was, on the dirt road along the creek. Among the building highlights are the 1890 schoolhouse (the oldest remaining structure), the workers' cottages (usually, two families lived in one small hut), the "newer" sawmill next to the creek (an eighteenth-century sawmill is being restored), and the imposing sorting and packing house.

Just below the packing house you'll see the beginning of the well-marked one-and-a-half-mile nature trail. This very picturesque sandy trail takes you on a loop along a series of cranberry bogs (now under water or under marshy growth), which you can easily identify with your nature guide. Fresh, clean water is everywhere, some areas deep, others quite shallow, all carefully dammed and controlled. (Fishing is allowed in some areas.) If you visit in late September or early October, you might see workers actually harvesting the cranberries. When shaken from the vine, the berries float to the water surface, and then they are gathered by machines. During the winter months the bogs are flooded to protect the plants from the bitter cold.

The cranberry theme continues at Pakim Pond within Brendan T. Byrne State Forest (Route 72). Cranberry bogs existed here, too, as they did in much of the region, and the pond was once a reservoir used to store water for the fall flooding of a nearby bog (now a swamp). The word *pakim* is, in fact, a Leni-Lenape Indian term for "cranberry."

A one-mile nature trail, which loops around the pond and swamp, is a pleasant, easy, woodsy walk on beds of soft pine needles. A self-guided tour (available at the park office) notes several points of interest where you can spot many varieties of plants and animals, from carpenter frogs, red-bellied turtles, raccoons, foxes, opossums, water snakes, and myriad birds to carnivorous plants, arrowhead plants (once used by Native Americans), and, yes, wild cranberries, sweet bay magnolia, and the prolific mountain laurel. You'll notice quite a number of dead trees (where migratory birds like to nest), drowned from too much water.

Within the much larger Brendan T. Byrne State Forest, the cranberry industry is represented by another intriguing place, Whitesbog Village. The village, listed as both a National and State Historic Site, is mostly intact, with many of its quaint old buildings from the early 1900s still standing. Surrounding it are the remains of what was once one of the largest cranberry and blueberry farms in all of New Jersey, originally owned by the White family.

Before starting off, be sure to stop at the Company Store or the kiosk on the edge of town, where maps and self guides are usually available. After poking around the village (part of which is being restored) you can take two walking trails: the Old Bog Nature Trail (a short one through the edge of an old abandoned bog) and the Whitesbog Village Tree Trail (where you'll see a good sampling of tree varieties found in the area). If you're up for a rare adventure, we recommend the five-mile self-guided driving tour that loops from the village through the nearby cranberry bogs, blueberry fields, and pine forests and across dams and canals on a network of narrow sandy roads. The fascinating vistas range from villagey to woodsy to vast moonscape-like panoramas with dead tree stumps in water, sand flats, and bright-orange sand paths. You'll find the setting surprisingly desolate—we met not a soul—but truly mesmerizing.

70 ·
HISTORIC FARMSTEAD GARDENS
WITH AN ORCHARD AND
HERB GARDEN

The Pleasures of Barclay Farmstead, Cherry Hill

❧ DIRECTIONS

209 Barclay Lane, Cherry Hill, NJ. Take the New Jersey Turnpike to exit 4 and go north on Route 73 to I-295 south. Take exit 34 off I-295 onto Route 70 west. Make a left on West Gate Drive and bear left onto Barclay Lane.

❧ INFORMATION

Gardens open daily, dawn to dusk. (House open Tuesday–Friday, 9 A.M.–3 P.M.) Telephone: 856-795-6225. Web site: www.barclayfarmstead.org.

❧ The Barclay Farmstead, with its thirty-two acres of apple orchards, old farmhouse, herb garden, and aura of a time long gone, is a lovely place to walk. You can still imagine, as you walk through the grounds—particularly in spring, when the trees are blossoming—just how appealing the place looked in the early 1800s. It has been beautifully restored and offers local residents a place to grow vegetables and flowers in an otherwise increasingly developed area.

The farm was first owned by a British family named Kay. In 1816 it was purchased by Joseph Thorn, who apparently constructed the eleven-room house (now used for exhibitions and events). In 1826 the property was bought by a founder of Camden, Joseph Cooper, after whom its apples—Cooper's Russetings—were named. Among the various Cooper descendants were the Barclays, and they lived on the farm until the 1950s. Today, despite the surrounding development, the farmstead is still very inviting.

In addition to the orchard, there is also a brief woodland trail on the property and an herb garden to enjoy, particularly in summertime. But if walking among apple blossoms is your pleasure, take this walk in springtime.

71 ·
FROM ARUGULA TO ZUCCHINI

A-to-Z Farm Offerings at Festive Terhune Orchards,
Princeton

✐ DIRECTIONS

330 Cold Soil Road, Princeton, NJ. Take the New Jersey Turnpike to exit 9, to Route 1 south, and continue for about twenty miles. Take I-95 south to exit 7B, take Route 206 north (Lawrenceville/ Princeton) to the fourth traffic light, and turn left onto Cold Soil Road. The main farm is on the right; the pick-your-own apple farm is around the corner on the right, on Van Kirk Road.

✐ INFORMATION

These two farms are under the same management. Call for information on what is available (the Van Kirk farm has apples and raspberries, and the main farm has extensive produce—to pick or buy). Open during the entire growing season. The farm offers many special events, including a Blueberry Bash (in July), Apple Days (in September), a Kite Day (in May at a ten-acre pasture), pumpkin picking, farm and nature outings, wagon rides, and pony rides, to name just a few. Telephone: 609-924-2310. Web site: www.terhuneorchards.com.

✐ This two-hundred-acre farm and orchard literally offers something for everyone. The owners have turned the old farm into a bustling enterprise, with thirty-five different varieties of fruits and vegetables, which you can pick or buy or just walk through. Produce grown on the farm is available from May through December. The pick-your-own produce includes berries of all kinds, in addition to peaches, cherries, pumpkins, and, of course, many different types of apples. This is a child-friendly spot and is a good way to spend the day with the entire family.

72·
THREE MUSHROOM FORAYS
INTO THE FOREST

*Cheesequake State Park, Matawan; Jenny Jump
State Forest, Hope; and Stokes State Forest, Sandyston*

❦ DIRECTIONS

Cheesequake State Park: *300 Gordon Road, Matawan, NJ. Take the Garden State Parkway south to exit 120, toward Laurence Harbor/Matawan. Make three right turns: Matawan Road, Cliffwood Road, and Gordon Road. Gordon Road goes right to the entrance.*

Jenny Jump State Forest: *330 State Park Road, Hope NJ. Take I-80 west to exit 12, Route 521. Take Route 521 south into Hope and make a left at the light onto Route 519, a right on Shiloh Road, and a left on State Park Road to the park.*

Stokes State Forest: *1 Coursen Road, Sandyston, NJ. Take I-80 west to exit 25, Route 206 north, This road will take you directly to the parklands.*

❦ INFORMATION

For information on the New Jersey Mycological Society: 908-362-7109 or www.njmyco .org.

❦ While mushroom walks have long been popular in Europe, they are a relatively recent U.S. discovery. The New Jersey Mycological Society has hundreds of members, and their outings are well attended. Among their many spots for hunting mushrooms throughout the state are Cheesequake State Park, Jenny Jump State Forest, and Stokes State Forest. The society organizes outings to these three locations, among many others, and we encourage you to go on a congenial walk with an expert in attendance. (See sidebar "Picking the Right Mushroom" for information.) While there are more than thirty thousand varieties of mushrooms, only a few hundred are known edibles, some of which are gourmet delights. About two dozen are deadly poisonous, and the society's newsletter often describes terrible errors made by those who do not know one mushroom from another. For this reason, and because the society has all the specifics on what to look for and where, we recommend that you join up with it (either as a guest or a member) for a mushrooming foray.

The outings of the Mycological Society are listed in advance. You can find out about them at www.njmyco.org. The society is a nonprofit organization that goes on regular forays under experienced leaders on Saturday or Sunday mornings (usually starting at 10 A.M.), from early May to late October. Among

Picking the Right Mushroom

Even if you are out foraging for mushrooms with a group of experts, you might like to know a thing or two about recognizing a few that grow in the region.

The Maitake: The "Hen of the Woods" or the "Dancing Mushroom" is highly regarded in Asian cuisine for its rich flavor and health benefits. You'll spot it as a whitish-gray or tan mushroom that grows in clusters and can often be found at the base of oak trees.

The Oyster Mushroom: This is a year-round mushroom that is white and mild in flavor and has a fragrance reminiscent of anise. (And it looks like an oyster!)

The Morel: This is one of the most common mushrooms around the region. Characterized by a brownish, spongelike top, the morel is prized in some countries and not in others. (Some people recommend boiling it and discarding the water.)

The Chanterelle: A yellow mushroom with a pleasant fruity odor, the chanterelle is a choice edible (particularly in European cuisine).

The Puffball Mushroom: Native to this area, the puffball can grow to be huge (as big as a soccer ball, we've heard!). They should be picked when the insides are thoroughly white. (Don't be confused with a false puffball, which is smaller and has dark spores.)

The Chicken Mushroom: This mushroom gets its name from its texture, which is similar to cooked white-meat chicken. It has a yellow-orange body and is a vegetarian favorite.

If you want to go mushroom picking, contact one of the following Mycological Societies: New York, www.newyorkmyc.org; Connecticut/Westchester, www.comafungi.org; Long Island, www.limyco.org; Mid-Hudson, www.mushroomthejournal.com; New Jersey, www.njmyco.org.

the delights of a walk with the group are the beautiful locations of these three natural settings, as well as the elaborate mushrooms-filled meals the society dishes up at various times. Joint trips with other mycological societies are also arranged in surrounding areas and farther afield, including Massachusetts, the Adirondacks, Long Island, and even abroad. Some of these distant outings last for several days.

The mushroom forays are held rain or shine, so be sure to dress accordingly. You should take paper bags (never plastic, which can damage the mushrooms) or waxed paper, a knife, and a basket. Hiking shoes and mosquito repellant are recommended. You should also take a picnic lunch. The leader will explain to you the proper system for cutting, identifying, sorting, and preserving your mushrooms. Collectors are encouraged to bring cameras and/or sketching materials to record their finds. The society will help you with identification charts and other material. Do not bring alcohol or dogs.

We recommend this sort of excursion primarily to adults. Both Stokes and Jenny Jump are quite hilly, so you should be comparatively energetic and sure-footed, but the mushroomers take their time as they hunt. This is not a walk to hurry through! The best season for mushrooming is late summer and fall. The right time and location for the prized morel, for example, is researched by the organization, and trips are planned accordingly.

We want to reiterate the advice of the experts: don't pick and eat any mushroom unless you have the advice of a knowledgeable guide! These are among the few walks that we don't encourage you to do on your own.

The Art of Foraging

Although people have long foraged for mushrooms, the newest trend in chic restaurants from coast to coast is using foraged ingredients, such as ginger root, sumac fruit, wild asparagus, lichen, Russian olive berries, licorice fern, pineapple weed, wild watercress, pine needles, radish flowers, nasturtiums, and mustard flowers. (Search "foraging" on the Internet, if the subject whets your appetite.) Some of these odd ingredients can be used in pasta dishes or drizzled over fish, for example. But like mushrooms, safe foraging is tricky. It's best to try it with an expert!

A book devoted to culinary artisans, including foraging, is *Food Heroes* by Georgia Pellegrini. For more information on foraging, contact Marion Bush, a forager of local wild edibles in the lower Hudson Valley, at marion_bush@yahoo.com.

73·
A PICTURE-BOOK ORCHARD
An Outing to Longmeadow Farm, Hope

❧ DIRECTIONS

561 Blairstown Road, Hope, NJ. Take I-80 to exit 12, Route 521. Go north one mile toward Blairstown; you'll find Longmeadow Farm on your left.

❧ INFORMATION

Open weekends, 9 A.M.–5 P.M. Best to visit in fall for apple harvest, the first week in August for raspberries, or the first week in October for pumpkins. Our recommendation is mid-September–October. The farm is highly recommended for a visit with children. Telephone: 908-459-5351. Web site: www.longmeadowfarmnj.com or www.upickapples.net.

❧ If you imagine a child's drawing of apple trees covered with big red circles, you won't be far off in picturing the glorious bright-colored apples that fill these trees in astonishing profusion. This delightful spot is a casual, open place that welcomes visitors who either pick apples (or pumpkins or raspberries or blackberries) or just walk around enjoying the atmosphere. It is definitely the

Longmeadow Farm, Hope

friendliest and most picturesque of the many orchards we have explored (accounting perhaps for the twenty-five thousand visitors a year).

Longmeadow Farm grows fourteen varieties of apples on its thirty-five acres of gently hilly terrain. The growers obviously have an extraordinary system of ecologically sound cultivation. There are nice dirt paths to walk among these incredibly profuse trees. The apples range from Ginger Gold to Macoun and McIntosh, Red Delicious to Fuji and Northern Spy. Any apple you pick will be a treat. (The orchard gives out recipes for apple use, as well as a guide to the different varieties.)

You might wonder how this number of apples will ever be used up. One answer is that in addition to providing fruit to the many you-pick-it visitors, Longmeadow Farm donates bushels of apples to the local food bank.

Whether you go to pick or to wander, we think you will agree with our assessment: Longmeadow Farm is a great example of nature's bounty.

74·
CHERRIES, PEACHES, AND APPLES
Take the Family to Battleview Orchards, Freehold

✒ DIRECTIONS

92 Wemrock Road, Freehold, NJ. Take the Garden State Parkway to exit 123, Route 9 south. Go fifteen miles to Route 33 west. Go about one mile, and at the second light make a right onto Wemrock Road; the orchard is about fifteen hundred feet on the right.

✒ INFORMATION

Open daily, in season; call for specific picking times. Telephone: 908-859-0599. Web site: www.battlevieworchards.com.

✒ Just next to the historic Monmouth Battlefield is this vast and inviting orchard: one hundred acres of neat rows of apple, peach, and cherry trees—and in fall, a wonderful selection of pumpkins (thousands of them!). Pick your own, or wander around at will in season. If you've never walked in an extensive orchard, you'll find it a particular pleasure; you feel a bit like a chess piece on a vast, orderly board!

75.

WINE, ECOLOGY, AND GLAMOUR

Laurita Winery, New Egypt

℘ DIRECTIONS

35 Archertown Road, New Egypt, NJ. Take the New Jersey Turnpike to exit 7A, then take I-195 east to exit 7. Take Route 526 east (toward Allentown), and after about one mile turn right at the stop sign onto Route 539 (Main Street); go through town and make a sharp left on High Street (still Route 539). After three miles, bear right (still Route 539); make a right turn on West Colliers Mills Road and continue onto Archertown Road. The winery is on your right.

℘ INFORMATION

Open year-round, Monday, Thursday, and Friday, noon–dusk; Saturday–Sunday, 11:30 A.M.–dusk. Closed Tuesday and Wednesday. Telephone: 609-758-8000. Web site: www .lauritawinery.com.

℘ For a one-of-a-kind vineyard experience we invite you to visit Laurita Winery. You would never expect so glamorous a site in this otherwise rustic area. When you first discover it, you think you've stepped onto a Hollywood movie set! Before you is a vast and impeccably groomed property surrounding an imposing modern structure (the three-story winery). Everything has been carefully designed, from the decorative plantings and stone walkways in front of the winery to the enormous deck overlooking rows upon rows of immaculate vines (with thirty-six thousand plants!). Piped-in music (Tony Bennett style) greets you as you walk up to the house, an unexpected touch we had not experienced before in other vineyards.

But Laurita is not just a pretty face. In fact, this eco-friendly site is a state-of-the-art sustainable operation with emphasis on conservation of resources. An impressive percentage of the winery structure—both inside and out—is reclaimed and recycled, as is the asphalt used in floors and parking areas; solar

---

For the Do-It-Yourself Winemaker

Have you always wanted to make your own wine? Apparently you are not alone. Make Wine with Us is an organization that offers a nine-session course, which begins with choosing the grapes and crushing them and moves on to pressing, bottling, and labeling and, finally, to a wine tasting. Telephone: 201-876-WINE. Web site: www.makewinewithus.com.

panels produce the needed energy for wine production; and the landscaping itself is designed to retain and conserve water. When you ask for a brochure of the vineyard, you are told that to save paper none has recently been printed; instead you are urged to check Laurita's Web site for additional information. (The friendly staff is also happy to answer further questions.)

Laurita Winery is a relative newcomer to New Jersey's steadily growing wine-production scene. Dating from 1998, when owners Randy Johnson and Ray Shea planted the first twenty acres of chardonnay, it now produces almost sixteen thousand cases of estate grown and bottled wine, including its highly regarded cabernet, merlot, and pinot gris, among others. Some 40 acres (of the 250-acre property) are dedicated to the vineyard itself. A few more acres are being used to test grapes that are not usually grown in New Jersey—a relatively recent experiment. Laurita's philosophy on grape growing and winemaking is very much hands-on. They suggest, "Carefully observe the grapes while they ripen," then "concentrate on turning what nature has provided into wine that everyone will enjoy."

This is a wonderful place to visit, whether you are a wine connoisseur or not. Not surprisingly, you are likely to find quite a number of enthusiastic weekenders, looking for a romantic getaway. Walking among the rows of vines is a real pleasure here, as is exploring the grounds, taking in the spectacular views from the deck (while sipping some wine or enjoying a picnic lunch), and discovering Laurita's many offerings. There are vineyard tours (on which you can learn a lot about wine production), three wine-tasting bars, an enormous dining and entertainment hall, a gift shop, and a marketplace with gourmet items. Right next to the vineyard is the Laurita Equestrian Center, as well as a charming, recently restored B&B (Dancer Farm Bed and Breakfast Inn) and a 250-acre working farm complete with nature trails and wildflower meadows.

76·
TWO VAST FARMS WITH
ETHNIC VEGETABLES

DeWolf's Farm and Hallock's U-Pick Farm,
New Egypt

ᴪ DIRECTIONS

DeWolf's Farm: *58 West Colliers Mills Road, New Egypt, NJ. Take the New Jersey Turnpike to exit 7 and take Route 528 south to New Egypt. Turn right on Route 539 to West Colliers Mills Road, and then turn right again to the farm.*

Hallock's U-Pick Farm: *38 Fischer Road, New Egypt, NJ. Take the New Jersey Turnpike to exit 7 and take Route 528 south to New Egypt. You'll find the farm on your right at the intersection with Fischer Road.*

ᴪ INFORMATION

DeWolf's Farm: Open daily, weather permitting, spring and fall, 7 A.M.–4 P.M.; summer, 7 A.M.–7 P.M. Call for crops available; you can pick your own or purchase them at Kim's Country Store on the farm. Telephone: 609-758-2424. E-mail: DeWolfarm@ aol.com

Hallock's U-Pick Farm: Open daily, weather permitting. Call for hours and crops available. Telephone: 609-758-8847. Web site: www.hallocksupick.com.

ᴪ You don't expect to see farmworkers carrying baskets of produce on their heads in this part of the world, but at Hallock's U-Pick Farm, which has many Jamaican workers—and island vegetables—that's part of the scene. At DeWolf's Farm you can pick or buy Korean produce. At both farms you can either drive or walk along the dusty farm roads through the huge fields, noting the signs for pickers of what's available. Ethnic vegetables abound: there are Thai peppers, Jamaican sweet potatoes, jalapeño peppers, Korean cabbage, and so on. The great variety of ethnic vegetables will delight those who enjoy foreign cuisines. In the huge, flat farmland areas of central New Jersey, these two large (and near each other) farms are well worth a visit. (You can pick your own or buy produce from their stands.)

At DeWolf's Farm there are all kinds of farmy things for the family to do: hayrides, tractor rides, pumpkin picking, a corn maze, and other such enjoyments. There are also flowers, as well as vegetables; it was the vast. more than two-hundred-acre panoply of vegetables that we enjoyed on our walk through the fields. You'll see Korean vegetables and bitter melons, cilantro, Asian long beans, sweet potato leaves, okra, African eggplants, blackberries, and many more types of fruits and vegetables growing here. You can pick peanuts (which are roasted at the shop).

Hallock's U-Pick Farm specializes in Jamaican (and American too, of course) vegetables; here we felt as if we were in a Caribbean landscape. The farm provides you with a list of produce to pick at any given time (bring your own knife!), including many kinds of peppers (try the Jamaican hot peppers), kohlrabi, sweet potato leaves, jute leaves, pumpkins, rutabagas, and much, much more. There are also sixty thousand feet of greenhouses, which are open from March to July; and the farm provides all things related to gardening and flower growing, and, yes, it has canning supplies.

77 ·

A HIKE (OR A HAYRIDE) THROUGH A VAST FARM

Peaceful Valley Orchards, Pittstown

✿ DIRECTIONS

150 Pittstown Road, Pittstown, NJ. Take the New Jersey Turnpike to I-78 west, to exit 15. At the bottom of the exit ramp turn left onto Route 513 (also called Pittstown Road) and continue for a mile and three-quarters. The entrance is on your right.

✿ INFORMATION

Open in growing seasons. Best for apple picking in fall and for other produce in summer and fall. Telephone: 908-730-7748 or 908-713-1705. E-mail: fruit@peacefulvalley orchards.com.

✿ You probably don't think of farms as hiking venues, but this one is an ideal spot for a most enjoyable long walk. Peaceful Valley's 150 acres are criss-crossed with pleasant dirt roads (which farm vehicles use), and you can walk into many different areas—from orchards to corn maze, flower beds to vegetable patches—or you can end up on a hayride.

This is a large site (for this area of the country), with a friendly feel and lots to attract families with children. But we also found its open spaciousness appealing: an invitation to walkers with an ever-changing scenery alongside, as red pepper beds give way to melon patches, tomatoes to pear trees—you get the idea!

You can, of course, pick your own here: huge orange pumpkins lying on the ground in October, peaches and pears in profusion on neat rows of trees in summer and autumn, some eighteen kinds of apples (all identified), cantaloupes, rhubarb, berries galore. (Or you can buy already-picked produce at the market.) And if you bring the kids, they will particularly love the corn maze with its unusual tunnel—an addition we hadn't seen before. There is also a summertime farm camp.

This farm is not a great distance from the urban areas to its east, but it seems like a world away, as you walk through the heady scents and visual treats of its thriving and diverse rows of growing things. The young couple who met in agriculture school and bought this farm less than a decade ago have achieved a family-friendly way to run a successful farm/orchard, and we thoroughly enjoyed our visit.

78 ·

"THE DEVIL'S TONGUE"
AND OTHER PEPPERS

Nearing Five Hundred Types of Chile at Cross Country Nurseries, Rosemont

❧ DIRECTIONS

199 Kingwood-Locktown Road, Rosemont, NJ. Take the New Jersey Turnpike to I-78 west, to exit 15. At the bottom of the exit ramp turn left onto Route 513 (also called Pittstown Road). Continue straight (no turns) for nine and a half miles. At the stop sign turn right onto Route 12 west. At the light turn left onto Route 519 south. Go about four miles and turn left onto Kingwood-Locktown Road. Nursery is a third of a mile on the right.

❧ INFORMATION

Open daily, weekdays, 9 A.M.–5 P.M.; weekends, 10 A.M.–6 P.M. Season begins in April and runs through early June. Call for opening day. The best time for visitors is mid-April through the end of May. Telephone: 908-996-4638. Web site: www.chileplants .com.

❧ You may have thought chile peppers were a product of the Southwest, but in fact, here in Rosemont, New Jersey, you'll find the largest growers of chiles in the world. This unusual enterprise adds new varieties all the time; it is near-ing the astonishing number of five hundred types of chile peppers! This is a very big enterprise: it turns out that chiles come in many different shapes and sizes, and range in taste from sweet to spicy to fiery hot.

The owners have been specializing in chile peppers since 1993. Here at Cross Country Nurseries you can buy the different plants ready to grow at home or as pods. The plants as seeds are started on heating mats under lights, beginning in January; after three weeks they are transferred to a greenhouse. By April they are ready for transplanting. Those that are to be grown outdoors need warmth (May or June in this area). In the outdoor growing fields some seventy-five of the most popular chiles are grown, while the most exotic of the many types of chiles are produced in a special greenhouse called the "motherhouse."

All the growing here is done with organic methods. Plants are fertilized with fish emulsion and seaweed. Ladybugs are released into the greenhouses, in or-der to eat aphids and other insect pests. A small pond in the motherhouse keeps the air moist and repels spider mites.

The chiles come from all over the world, and the names of the many va-rieties indicate the widespread taste for this product: Vietnamese Multi-Color,

Zimbabwe Bud, Bharta Hybrid, and Bhut Jolokia, among many others. But the most commonly chosen by purchasers are The Devil's Tongue, Red Savino Habañeros, African Fatalli, and Chocolate Habañero. The owners say that the hottest chiles are the most popular!

If you like chiles, or even if you're just interested in this most unusual enterprise, don't miss a visit here.

AND IN KEEP IN MIND . . .

79 • Fruitwood Orchards Honey Farm, Monroeville

ADDRESS: *419 Elk Road, Monroeville, NJ.*
INFORMATION: Telephone: 856-881-7748. Web site: www.fruitwoodorchardshoney.com.

This honey farm in Gloucester County also has a variety of fruits on its 110 acres, where you can pick peaches, apples, and so on. You can visit the hives and see the (organic) process of honey making. Call ahead of time for seasonal information.

Grapes ready for harvest (iStockphoto.com/Dwight Nadig)

80 ◆ Westfall Winery, Montague

ADDRESS: *141 Clove Road, Montague, NJ.*
INFORMATION: Telephone: 973-293-3428. Web site: www.westfallwinery.com.

Not far from the Delaware Water Gap in the hills of Sussex County is this picturesque vineyard and winery. The vineyard covers 6 acres (of a 338-acre property), and you can walk around (and bring your pets). You can picnic, enjoy a winetrail weekend, and experience wine tastings (including warm apple-cinnamon wine), as well as tasting wine at various stages of the aging process. The farm has remained in the same family since 1774. Among the wines are sangria, cabernet sauvignon, peach chardonnay, and a Spanish Passion.

81 ◆ Stults Farm, Cranbury

ADDRESS: *62 John White Road, Cranbury, NJ.*
INFORMATION: Telephone: 609-799-2523. Web site: www.stultsfarm.com.

This large farm has been in operation since 1915. Its two hundred acres include you-pick-it vegetables and fruits on preserved land. Among the offerings are various Asian specialties (Indian and Chinese bitter melon and Asian pears), as well as berries, squashes, peppers, eggplants, and so on.

PENNSYLVANIA
AND DELAWARE

82·
HIKING A MEDICINAL TRAIL

Bowman's Hill Wildflower Preserve,
New Hope, Pennsylvania

🐾 DIRECTIONS

1635 River Road, New Hope, PA. Take the New Jersey Turnpike south to I-195 west. Continue on Route 29 north for about twelve miles and make a left at General Washington Memorial Boulevard to go over the bridge. Turn right onto Route 32. Bowman's Hill is about five miles northwest on Route 32.

🐾 INFORMATION

Open year-round except for major holidays. Pick up a flyer advising which path to take. Wildflowers are best in spring and early summer, but there is something to be seen in any season. One trail is wheelchair accessible. Telephone: 215-862-2924. Web site: www.bhwp.org.

🐾 As you walk along the woodsy trails and gentle creek in this lovely spot, you'll feel you are miles from the bustle of the highway and civilization. The preserve includes in its one hundred acres quite a variety of habitats: forest, meadows, ponds, bogs, an arboretum, and preserves for shrubs and flowers. An azalea trail leads to a path of bluebells; a walk brings you to marsh marigold and holly areas; you pass the charming dam on the Pidcock Creek and come to the evergreen area. More energetic visitors can hike (or drive) to the famous Bowman's Hill Tower at the top of a surprisingly high hill in the preserve and thus complete their visit to this historic and lovely place. Your choices are extensive. This garden walk is hilly, but the preserve has more gentle trails as well. We recommend this outing to anyone who enjoys wildflowers, woods, and hiking. Children especially will enjoy the chance to roam freely.

Established to preserve Pennsylvania's native plants, Bowman's Hill opened in 1934. It is made up of twenty-six trails, including a trail for the handicapped (wheelchair accessible) and a famous arboretum called Penn's Woods, which has more than 450 different trees and shrubs. The general atmosphere of Bowman's Hill is relaxed. It is rather like wandering through a large, private nineteenth-century estate.

Among these numerous trails here at Bowman's Hill is a most unusual one: the Medicinal Trail. It is one of the most exceptional and educational parts of the preserve and is devoted to plants with medicinal uses, both real and legendary. The Medicinal Trail is a special walk for herb fanciers and those who are interested in natural cures of Native American herbal medicines. It includes plants of scientifically proven effectiveness as well as plants whose curative properties are nothing more than old wives' tales. A descriptive flyer tells you what you're seeing and what it's good for—or what legends say it's good for.

The trail, which is about 620 feet long, is slow going if you stop to examine each plant. While you are warned not to taste anything yourself—some of the plants are extremely poisonous—you are encouraged to go slowly and study each plant's story. Due to the dangerous nature of many of the plants, this trail is not recommended for small children.

The trail begins with mountain laurel, the juices of which the Indians supposedly drank to commit suicide (although by 1800 a tincture made from its leaves was being used to treat several diseases). Next comes mayapple and bloodroot, a source of morphine that Indians chewed to cure a sore throat. Other plants include white oak, spicebush, wild ginger, alumroot, fairywand, and witch hazel. Among the old wives' varieties is ginseng, which Europeans thought restored youth (and which has been used for centuries as a cure-all in China). The list is a long one, and a walk among these plants is extremely interesting.

Other walks in Bowman's Hill Wildflower Preserve can of course be combined with this walk on the same day. There are a tree identification trail, wildflower trails, and several kinds of nature walks in this preserve. Several times during the year the preserve offers guided mushroom walks, bird walks, native plants and tern identification walks, as well as other wild-plant walks. Each season brings different plants and appropriate events.

If you go downhill from the car park, you will come to the creek, which you can follow along its winding path. If you continue along the roadway (not accessible to cars) to its end, you will come to a rugged climb to Bowman's Hill Tower.

The hike up to the tower can be rigorous, and the trail takes you through dense underbrush. Your effort is well rewarded at the top. Bowman's Hill Tower stands on a hill 380 feet above sea level, one of the highest points along the Delaware River. The hill long served as a landmark, and before George Washington's crossing of the Delaware it was used as a lookout and signal station. The present tower commands a view of fourteen miles of the Delaware River Valley, including the very spot that Washington crossed. If you prefer to drive to the tower, you can get in your car and drive out of the preserve.

83·
FRUITS, LAVENDER, AND WINE

An Agrarian Visit to Bucks County's Solebury Orchards,
Carousel Farm Lavender, and Wycombe Vineyards,
New Hope Vicinity, Pennsylvania

❧ DIRECTIONS

Solebury Orchards: *3325 Creamery Road, New Hope, PA. Take the New Jersey Turnpike to I-78 west and take exit 29 to I-287 south. Take exit 17 onto Route 202 south. Go through the Delaware River tollbooth and take the first exit to your right. Turn left onto Route 32 north, go half a mile, and turn left onto Phillips Mill Road. Go about two miles to the stop sign at Route 263 and turn left onto Route 263 south. Go half a mile and make the first right onto Creamery Road.*

Carousel Farm Lavender: *5966 Mechanicsville Road, Mechanicsville, PA. Follow directions to Solebury Orchards; continue on Creamery Road for a short distance and make a left on Mechanicsville Road. Continue for about two miles and make a left on Sheffield Drive to the entrance.*

Wycombe Vineyards: *1391 Forest Grove Road, Furlong, PA. Follow directions to Carousel Farm Lavender but continue on Mechanicsville Road to the intersection with Route 413. Make a left onto Route 413 and continue through Buckingham; make a right on New Hope Road and a right at the T intersection (Forest Grove Road). The second driveway on your right is the vineyard.*

❧ INFORMATION

Solebury Orchards: Open in growing season, Thursday–Sunday, 9 A.M.–6 P.M. Telephone: 215-297-8079. Web site: www.soleburyorchards.com.

Carousel Farm Lavender: Open Saturdays only, 9 A.M.–5 P.M. Group tours by appointment. Telephone: 917-837-6903. Web site: www.carouselfarmlavender.com.

Wycombe Vineyards: Open Wednesday–Friday, noon–5 P.M.; Saturday, noon–6 P.M.; Sunday, noon–5 P.M. Telephone: 215-598-WINE. Web site: www.wycombevineyards .com.

❧ The Bucks County area is home to many beautiful farms, orchards, and vineyards. This rustic yet very elegant region has retained its bucolic charm, with lovely old stone houses and winding country roads, despite its proximity to urban sprawl. There is a surprising amount of unspoiled farmland with thriving agricultural enterprises. We found the folks we met were friendly and welcoming. We have chosen three of our favorite farm sites, all in close proximity to one another. (And you can enjoy a lunch nearby in charming New Hope on the Delaware River.)

Solebury Orchards is a picture-perfect farm: scenic, beautiful, vast, unspoiled, and filled with an abundance of luscious produce ready to be picked. This is a

Solebury Orchards, New Hope (Courtesy Solebury Orchards)

quiet and somewhat secluded spot, where you can savor nature's bounty in a peaceful setting. The ambiance is inviting (you feel welcome from the moment you step foot on the grounds), relaxed, and low-key.

The farm includes some seventy acres of fruits and vegetables (and glorious flowers, too), and you will love walking around here amid the fields of berries and orchards. Known for its high-quality fruits, it produces peaches, apples, blueberries, raspberries, cherries, plums, and more. In addition, apples are pressed here to make cider and applesauce (which you will want to sample).

In summer and fall you are welcome to pick your own produce: blueberries from June through August, raspberries from mid-June to early October, cherries in late June, blackberries from late July to September, cherry tomatoes from early August through September, and of course apples, from late August to early November. You can even gather your own flowers. (The lovely cutting garden with its waist-high flowers is right in the front where you can't miss it.) Picking here is easy even on a rainy day, as there are rows of mowed grass between the berry bushes and fruit trees and therefore no mud to contend with as you fill your basket. On weekends visitors can take a wagon ride through the orchards for apple picking. Not surprisingly, Solebury Orchards has a faithful clientele: people come back, time after time, and perhaps you will, too!

Carousel Farm Lavender is not far away. If you've never seen lavender growing, it is quite a sight: long rows of clumpy, feathery, pale-colored—and very fragrant—plants stretch out before you. You can walk among the rows, divided by stone walls, and breathe in the delicious aroma. This is a vast scene; we were surprised by how much lavender was growing here. Many products are made from the distillation of these flowers, including foods (teas and honey), candles, soaps, creams, and sachets.

Carousel Farm Lavender is one of the largest organic lavender farms on the East Coast. First established in 1748, this bucolic site has had many incarnations over the years: first a dairy farm, it became a horse farm and later even an exotic animal farm. The present owners (who bought it in the early 2000s) drew their inspiration from their travels through Provence in France, where they saw fields and fields of lavender. This place in Bucks County seemed ideal for lavender cultivation. And the fact that it came with its own eighteenth-century stone barn and fieldstone walls only added to its inviting setting.

Here at Carousel there are basically four varieties of lavender and about fifteen thousand plants, all carefully tended to by hand, from planting to pruning to harvesting. And despite the harsher winters here (as compared to those in the south of France), the plants all seem to be thriving. It is not unusual to find local painters and photographers occasionally wandering about the grounds, gathering inspiration.

The farm (open on Saturdays only) offers guided tours, but you are welcome to walk around on your own, too. (And bring the kids, as there are also animals on the farm—among them goats, llamas, and horses.)

Our third stop here in Bucks County is Wycombe Vineyards. Wycombe is a quiet, seventy-acre, family vineyard in the heart of the county. The rustic property, which has been in the same family since 1925, has been used since about 2000 for the cultivation of grapes: early on, Wycombe was producing sod,

Carousel Farm Lavender, Mechanicsville (Courtesy Carousel Farm Lavender; photo by Niko Christon)

which, with its aeration practices adds supplemental limestone, enriching the soil over the years, making it especially beneficial to growing grapes.

The vineyard uses artisanal methods in its wine production: the grapes are all grown right here and are handpicked, and the processing of the wine is done on-site, as well. When you buy the vineyard's pinot noir, chardonnay, Riesling, or cabernet sauvignon (among other varieties), you know they are absolutely genuine.

We found the feel of this place appealing and friendly. As we sat on the terrace (where wine tastings are offered) with the engaging and easygoing owner, we were told that visitors are more than welcome to explore the vineyard on their own. A walk around and about the rows of vines turns out to be about six miles, but you can do as much or as little as you like. Make sure to sample some of the vineyard's offerings!

For Extra Spice in Your Food, Add Edible Flowers

Edible flowers have been used for flavor and garnish in many cultures for centuries, so adding them to a salad or as a decoration on a festive platter is nothing new. Here are a few tips to keep in mind:

- Though many flowers are perfectly safe and tasty (and rich in vitamins, too), some are poisonous! Stick to your basic list of edibles (see the examples in the following list).
- If you're picking your own flowers, do so early in the day, when their flavor is at its peak. Choose flowers with larger petals (such as pansies). Do not select flowers that have been sprayed with pesticides, including those from a florist.

Here are some recommended plants with edible flowers:

Apple or crabapple	Lavender
Basil	Lemon and orange
Broccoli and cauliflower	Lilac
Chive	Mustard
Chrysanthemum	Nasturtium
Clover	Okra
Daisy	Pansy
Dill	Rosemary
Fennel	Thyme
Garden pea	Violet

84·
HERBS AND TEAS IN AN
ART-FILLED ENVIRONMENT

Plantings and Sculpture at Morris Arboretum,
Philadelphia, Pennsylvania

❧ DIRECTIONS

100 East Northwestern Avenue, Philadelphia, PA. Take the Pennsylvania Turnpike (I-276) to exit 333 (Plymouth Meeting/Norristown). Take the Germantown Pike east for about three miles and turn left on Northwestern Avenue. The Morris Arboretum entrance is about a quarter mile up, on the right.

❧ INFORMATION

Check for hours and best times to visit. Telephone: 215-247-5777. Web site: www.business-services.upenn.edu/arboretum.

❧ The Morris Arboretum is a vast (166 acres) and fascinating place, combining nature and art in an unusually felicitous way. Part of the University of Pennsylvania, it emphasizes trees, shrubs, and contemporary sculpture. But it also has a collection of small gardens, including flower beds, a fern garden, rock gardens, and, of particular interest to us, an herb and tea garden. Here you'll find many herbs, both for cooking and medicinal purposes, and hybrid teas—very unusual for this part of the country.

85·
AMERICA'S EARLIEST BOTANICAL GARDEN

Vegetables and Herbs at Bartram's Garden,
Philadelphia, Pennsylvania

℅ DIRECTIONS

Lindbergh Boulevard and South Fifty-fourth Street, Philadelphia, PA. Take the New Jersey Turnpike south to I-295 south and then take I-76 west (Schuylkill Expressway). After crossing the Walt Whitman Bridge, exit going west on Passyhunk Avenue. Turn right on South Sixty-first Street and then right again onto Lindbergh Boulevard north; the entrance to Bartram's Garden is on your right at the corner of Fifty-fourth Street.

℅ INFORMATION

Garden open daily, dawn to dusk. Tours may be scheduled from 9 A.M. to 5 P.M. daily. Admission is free. The house is also open to visitors; call for hours and fee. Picnics are allowed, and lunches can be ordered in advance. The garden offers numerous educational and horticultural events. Note that the grounds are hilly and have occasional rough footing. Telephone: 215-729-5281. Web site: www.bartramsgarden.org.

℅ This is indeed a historic garden; there are many colonial gardens in this region, but Bartram's is the first, and most important, American botanical garden. Today, it is still one of the most fascinating colonial sites you can visit; the plants, trees, herbs, vegetables—everything you see is either a descendant of Bartram's original specimens or an exact replica.

While many restored colonial gardens show us the orderly and useful aspects of early American gardening, John Bartram's extraordinary, less-than-orderly historic garden takes us on a flight of fancy. Located on the banks of the Schuylkill River on the outskirts of the city of Philadelphia, Bartram's Garden is a botanist's and a historian's delight.

On this twenty-seven-acre hillside, just behind the 1730 house, are hundreds of tree and plant specimens that Bartram brought to the site more than two hundred years ago—and that are still flourishing. The visitor gets a special sense of American history because the garden has been tended continuously and Bartram's design kept intact. Unlike many other colonial gardens, this pleasantly disordered place gives a sense of the excitement of settlers in a new world.

John Bartram was a Quaker Philadelphian who lived from 1699 to 1777. A farmer and botanist, he began the first thorough collection of native plants in the country. His garden was started in 1728 and increasingly became a passion, as he gathered plants, seeds, and specimens of nature's bounty on the American continent. The adventurous botanist traveled all over the eastern half of the continent (as far as the Ohio River to the west and Appalachian Georgia and Florida to the south), carrying specimens back to Philadelphia in an airtight ox bladder. With his son, William, who continued the collection well into the nineteenth century, he established a commercial nursery, supplying seeds and plants by mail to such great gardens as those at Mount Vernon and Monticello. Bartram also received specimens from all over the colonies, from the West Indies, and from botanists worldwide. His carefully cataloged (and illustrated) collection formed the first botanical garden in the country and included an astonishing four thousand species. The Bartrams also introduced some two hundred American plants to Europe. Internationally known and admired, the father and son presented in England such subsequent staples of their world as mountain laurel and sugar maple. Frequently Bartram put together for shipment abroad "five-guinea boxes" of the seeds of plants discovered on his travels.

When you visit the garden, you will see, just behind the house, Bartram's medicinal herb and vegetable gardens re-created today to show the plants that would have filled his eighteenth-century garden. Here are patience and comfrey, for example, herbs used in earlier times as medicines (comfrey root was boiled and used for treating coughs and intestinal disorders), as well as "heirloom" vegetables in raised beds. There are reportedly seventy-eight different vegetables and herbs in these beds; it is a truly fascinating garden. (You can get a guide to the uses of the herbs at the bookshop.)

You'll also find a fascinating collection of eighty-two types of shrubs and trees, about four hundred yards down the gentle hill toward the river. Here you'll find giant oaks (known as Bartram oaks) and the famous Franklin tree from Florida, now extinct in the wild. There are ginkgo trees, prickly ash (whose bark was known as a toothache remedy), indigo bush, witch hazel, and bald cypress interplanted with mountain laurel, cucumber magnolia (brought from the shores of Lake Ontario), and the Fraser magnolia (from the Great Smoky Mountains). You'll see a pawpaw tree (of which legend has Bartram sending the fruit and flowers back to England in a bottle of rum) and in springtime masses of flowering dogwood, azalea, and wisteria. At the very bottom of the hill you'll come upon old rocks and an evocative landing spot on the river.

Despite the occasionally weedy aspects of this garden—no doubt historically accurate!—you'll relish the sense of history and the mental image of the colonial Quakers arriving home from the American wilderness to plant ever more specimens of flowers and trees in their Philadelphia garden. If you find Bartram's Garden of interest, do not miss the collection of William's delicate and meticulous watercolors of plants or his interesting description of traveling and plant collecting, published in 1791. Information on these historic items and even tours re-creating the Bartrams' journeys is available at the headquarters of Bartram's Garden.

86·
WHEN PLANTS ARE USED
IN HEALING

The Landscape Arboretum at Temple University,
Ambler, Pennsylvania

❧ DIRECTIONS

580 Meetinghouse Road, Ambler, PA. Take the New Jersey Turnpike to exit 6 and take the Pennsylvania Turnpike (I-276) to exit 339 at the Fort Washington interchange. Go north on Route 309 for about three miles. Exit at Susquehanna Road and turn left. Go to the first light, and make a right onto East Butler Pike. Go about half a mile to Meetinghouse Road and turn right. (Go to the main administration building to get a parking permit.)

❧ INFORMATION

Grounds are open daily, 8 A.M.–dusk. Greenhouses are open Monday–Friday, 8:30 A.M.–4:30 P.M. (closed on holidays). A variety of tours and events are offered. Telephone: 267-468-8400. Web site: www.ambler.temple.edu/arboretum.

❧ This garden is almost exactly one hundred years old; in 1910 Jane Browne Haines purchased a seventy-one-acre farm to found a school of horticulture for women. Over the years this historic and forward-thinking school trained hundreds of women from both the United States and abroad in horticulture. And 116 more acres were added. In 1958 the school became part of Temple University and began giving degrees in horticulture and landscape architecture.

The use of plants in the healing process goes back, of course, to antiquity, but even today numerous herbs are used for medicinal purposes. At this site there are both plants for healing and a carefully designed garden for contemplation. These two different gardens both relate to healing: the Viola Anders Herb Garden, where culinary, dye, and medicinal herbs were installed in 1992, and the Ernesta Ballard Healing Garden. In the Herb Garden, a variety of herbs used for curing disease are grown (as well as numerous herbs used for food). And nearby at the Healing Garden you'll find a "contemplative garden" (originally part of a show called "Nature Nurtures"), where there are native plants, a labyrinth, and an unusually lovely atmosphere.

87·
ESPALIERED FRUIT TREES

A Picturesque French Tradition Revived at the
Gardens at Eleutherian Mills, Wilmington, Delaware

❧ DIRECTIONS

200 Hagley Creek Road, Wilmington, DE. Take the New Jersey Turnpike south across the Delaware Memorial Bridge and follow signs to Route 141 north. Go north for seven miles, and shortly after you cross Route 100 you will come to Hagley's main entrance on your left.

❧ INFORMATION

Open daily, 9:30 A.M.–4:30 P.M. (Check for special hours in winter.) Tours are available. There is an admission fee. Telephone: 302-658-2400. Web site: www.hagley.org.

❧ This is a vast, beautiful, and historic site to visit in so many ways: its founding and subsequent history as a major early American industrial site, its beautiful mansion, and, for us, its gardens. And what gardens they are! Ranging from rows of carefully alternating flowers and vegetables to exquisite rose gardens and honeysuckle-covered latticework, these are gardens to savor and to remember. And of particular interest are the fascinating espaliered fruit trees. These are a sight you won't want to miss.

First, a brief word about the past of this unusual, French-oriented site. Hagley, the 320-acre estate, with its elegant treasure- and antique-filled mansion, was the site of the gunpowder works established by the French-born industrialist E. I. du Pont. Its situation on a bluff overlooking a steep ravine along the Brandywine River provided both power for the industry and a glorious setting for the Georgian-style house and the exquisite formal gardens in the French Renaissance style. A visit and tour here will show you everything from how the water turbine worked to how the early American gardens (in historically accurate re-creations) looked and functioned. There are some sixty structures and more than two acres of gardens on the estate.

Eleutherian Mills was the name given to the family home. The original kitchen garden, which provided foods for the du Ponts and their seven children, has been re-created using original documents. The gardens were originally laid out between 1803 and 1834. Using materials from the du Pont archives, archaeologists were able to reconstruct the areas of gardens, paths, well site, summer house, and even the location of specific plants. Flowers, herbs, nut trees, vegetables, and fruit trees were planted—many in alternating rows of raised beds in the French style of geometric design. These gardens are thriving today:

Eleutherian Mills, Wilmington (Courtesy Eleutherian Mills, Hagley Museum)

vegetables range from cabbage and spinach to onions and broccoli; herbs include peppermint, summer savory, chamomile, and dill; fruits include apples, cherries, pears, strawberries, and peaches.

It is the fruit trees that particularly grabbed our attention. The French developed the espalier system in the Napoleonic era. It is essentially a way of saving space while providing a site for growing trees that would ordinarily require an orchard. Instead of the trees' usual setting, they grow along a wall or fence; they are grafted and pruned and shaped so that they appear almost two-dimensional against their backings. Both decorative and effective, the espaliered fruit trees here form what du Pont called "fences," and he used them to border growing beds as well as to produce apples, peaches, and pears for the estate.

Among the highly picturesque espaliers are the "pleached arbor" (interlaced branches) of lady apple trees, whose top branches intertwine over the pathway; pear trees espaliered into diamond shapes (called a Belgian fence), fan shapes, tent shapes, and U shapes; and even stunted trees that have been espaliered and trained into a ropelike form.

The system for espaliered fruit trees is a complex one involving grafting dwarf fruit trees onto hardier stock, cutting off branches to create a two-dimensional shape, and tying the remaining part of the tree into the desired shape for several weeks at a time. This very complicated and time-consuming method requires constant attention and trimming, and a master gardener to do it. (It is certainly a rarity in American gardens.)

You will enjoy this "reworking" of nature and the unusual patterns it creates. It reminds us of formal French garden traditions with their manicured trees and the art of topiary—another version of nature reimagined.

AND KEEP IN MIND . . .

88 ◆ Vineyards in the Delaware Valley

The picturesque Delaware Valley is also home to several vineyards that produce award-winning wines. Among them are Crossing Vineyards and Winery in Washington Crossing, Pennsylvania (www.crossingvineyards.com), which offers tasting and tours; Rose Bank Winery (www.rosebankwinery.com) in Newtown, Pennsylvania; and Sand Castle Winery in Erwinna, Pennsylvania (www.sandcastlewinery.com). Each specializes in different grapes, including pinot noir, cabernet sauvignon, and Riesling. Bucks County, Pennsylvania, with its cool sunny climate and tough rocky soil, has become an inviting wine region along with its other tourist attractions.

Gilbertie's Herb Gardens, Westport

CHOOSING AN OUTING

Farms

CORN MAZES

Lyman Orchards, Middlefields, CT: 28
Queens County Farm Museum, Floral Park,
  NY: 42
F&W Schmitt's Farm, Melville, NY: 65
Hacklebarney Farm, Chester, NJ: 143
Alstede Farms, Chester, NJ: 143
Peaceful Valley Orchards, Pittstown, NJ:
  164

DEMONSTRATION, EDUCATIONAL, AND CSA FARMS

Queens Botanical Garden, Flushing, NY:
  40
Biophilia Organic Farm, Jamesport, NY: 63
Quail Hill Farm, Amagansett, NY: 68
Stone Barns Center for Food and Agricul-
  ture, Pocantico Hills, NY: 71
Ryder Farm, Brewster, NY: 95
Cropsey Farm, New City, NY: 132

ETHNIC AND EXOTIC PRODUCE

Holbrook Farm, Bethel, CT: 19
Brooklyn Botanic Garden, Brooklyn, NY:
  34
New York Botanical Garden, Bronx, NY:
  36
Planting Fields Arboretum, Oyster Bay, NY:
  49
Hammond Museum and Japanese Stroll
  Garden, North Salem, NY: 86
Mianus River Gorge Wildlife Refuge and
  Botanical Preserve, Bedford, NY: 90
Ryder Farm, Brewster, NY: 95
Evergreen Farm, Hamilton Township, NJ:
  141
DeWolf's Farm, New Egypt, NJ: 162
Hallock's U-Pick Farm, New Egypt, NJ: 162
Stults Farm, Cranbury, NJ: 168

SCENIC FARMS

McLaughlin Vineyards, Sandy Hook, CT:
  11, 15
Hopkins Vineyard, Warren, CT: 15
Lasdon Park and Arboretum, Somers, NY:
  75
Stonecrop Gardens, Cold Spring, NY: 92
Dr. Davies Farm, Congers, NY: 107
Benmarl Winery, Marlboro, NY: 114
Whitecliff Vineyard and Winery, Gardiner,
  NY: 114
Laurita Winery, New Egypt, NJ: 160

YOU-PICK-IT FARMS, FRUIT

Blue Jay Orchards, Bethel, CT: 19
Wilkens Fruit and Fir Farm, Yorktown
  Heights, NY: 81
Stuart's Fruit Farm, Granite Springs, NY: 81
Fishkill Farms, Hopewell Junction, NY: 88
Greig Farm, Red Hook, NY: 104
Blueberry Park, Wingdale, NY: 104
Dr. Davies Farm, Congers, NY: 107
Orchards of Concklin, Pomona, NY: 113
Applewood Orchards, Warwick, NY: 121
Lawrence Farms, Newburgh, NY: 119
Warwick Valley Winery & Distillery, War-
  wick, NY: 121
Ochs Orchard, Warwick, NY: 121
Weed Orchards, Marlboro, NY: 130
Jenkins-Lueken Orchards, New Paltz, NY:
  134
Sussex County Strawberry Farm, Newton,
  NJ: 140
Alstede Farms, Chester, NJ: 143
Demarest Farms, Hillsdale, NJ: 147
Terhune Orchards, Princeton, NJ: 153
Battleview Orchards, Freehold, NJ: 159
Peaceful Valley Orchards, Pittstown, NJ: 164
Solebury Orchards, New Hope, PA: 173

YOU-PICK-IT FARMS, VEGETABLES

A VARIETY OF GROWING METHODS

Greenhouses, Arboretums, and Botanical Gardens

Herb Gardens

Historic Interest

Other Specialties

FOOD AND WINE FESTIVALS

Many of the farms and vineyards in this book sponsor or take part in festivals during the year. Here are a few that look inviting:

Annual Maple Sugaring Festival (March)
New Jersey Audubon Society
Weis Ecology Center, Ringwood, NJ
Telephone: 973-835-2160

White Silo Farm and Winery Annual
    Asparagus Festival (late spring)
Sherman, CT
Web site: www.whitesilowinery.com

HerbFest (June)
Pleasant View Farm, Somers, CT
Web site: www.ctherb.org

Dinners at the Farm (summer)
East Lyme, CT
Web site: www.dinnersatthefarm.com

Peas & Cukes Organic Gardening
    (summer)
Long Island, NY
Web site: www.motherhouse.us

Honey Weekend (July–August)
Wave Hill, Bronx, NY
Web site: www.wavehill.org

Wild Blueberry and Huckleberry Festival
    (July–August)
Ellenville, NY
Telephone: 845-647-4620

Tomato To-mah-to Heirloom Tasting Feast
    (August)
Cromwell, CT
E-mail: slowfoodct@earthlink.net

Tour of Easton Farm (August)
Easton area, CT
E-mail: jean@individualdifferences.com

Bounty of the Hudson Wine Festival (late
    summer)
Orange County, NY
Web site: www.shawangunkwinetrail.com

Alstede Farms' Fall Harvest Celebration
    (September)
Chester, NJ
Web site: www.alstedefarms.com

Apple Festival (September)
Esopus, NY
Web site: www.gbgm-umc.org/TOEUMC/
    applefestival.html

Kennett Square Mushroom Festival
    (September)
Brandywine Valley, PA
Web site: www.mushroomfestival.org

Wine for Beginners at the Wine Institute
    (September)
Crossing Vineyards and Winery, PA
Web site: www.crossingvineyards.com

Annual Cranberry Festival (fall)
Bordentown, NJ
Web site: www.downtownbordentown.com

Fall Blues and Pumpkin Festival (fall)
Milford, NJ
Web site: www.albavineyard.com

Annual Chili Day (October)
Bethel, CT
Web site: www.bethelwoodscenter.org

Apple Festival (October)
Queens County Farm Museum, Floral Park,
    NY
Web site: www.queensfarm.org

Connecticut Garlic and Harvest Festival
(October)
Bethlehem, CT
Web site: www.garlicfestct.com

Haight-Brown Vineyard Harvest Festival
(October)
Litchfield, CT
Web site: www.haightvineyards.com

Hopkins Vineyard Annual Harvest
Celebration (October)
Warren, CT
Web site: www.hopkinsvineyard.com

Hudson Valley Garlic Festival
(October)
Saugerties, NY
Web site: www.hvgf.org

Edible Sculpture Party

Of all the wonderful festivals we have discovered in this region, none is more intriguing than the Edible Sculpture Party held each year in Tivoli, New York. This event, thrown by two artists, requires you to contribute your own edible sculpture in order to attend—and it is well worth the effort! For here in July you'll find the most amazing concoctions: a statue suggesting a water tower made of carrots and eggplant, for example, or zucchini figures with tomato heads, as well as abstract shapes and designs of every sort. And when the exhibition phase is over, the sculpture is enthusiastically eaten by the visitors!

INDEX OF PLACES

ABOUT THE AUTHORS

Lucy D. Rosenfeld and Marina Harrison are the coauthors of nine guidebooks, including *Architecture Walks: The Best Outings near New York City* and *A Guide to Green New Jersey*. Among their books are walking tours of gardens, historic sites, public art, and nature preserves. They are lifelong friends who enjoy exploring new places and introducing them to their readers. Lucy D. Rosenfeld is also the author of more than twenty books on art and architecture. Marina Harrison, in addition to writing guidebooks, worked for many years in publishing.